Atlas of
VIRUS DIAGRAMS

Hans-W. Ackermann and Laurent Berthiaume

CRC Press

Boca Raton   New York   London   Tokyo

Cover diagram from J.C. Brown, and W.W. Newcomb, Rhabdoviridae, in Animal Virus Structure, M.V. Nermut and A.C. Steven, Eds., 1987, p. 199. Courtesy of Elsevier, Amsterdam. With permission.

Library of Congress Cataloging-in-Publication Data

Ackermann, Hans-Wolfgang, 1936–
    Atlas of virus diagrams / Hans-Wolfgang Ackermann, Laurent
Berthiaume.
       p.   cm.
    Includes bibliographical references and index.
    ISBN 0-8493-2457-2
    I. Berthiaume, Laurent. II. Title.
    [DNLM: 1. Viruses--atlases. 2. Viruses--classification. QW 17
A182a 1995]
QR363. A25 1995
576¢.64¢0222--dc20

DNLM/DLC
for Library of Congress
                                                 94-49598
                                                  CIP

No claim to original U.S. Government works
International Standard Book Number 0-8493-2457-2
Library of Congress Card Number 94-49598
Printed in the United States of America 1 2 3 4 5 6 7 8 9 0
Printed on acid-free paper

THE AUTHORS

Hans-Wolfgang Ackermann, M.D., is Professor of Microbiology at the Medical School of Laval University, Quebec. He was born in Berlin in 1936 and obtained his medical degree in 1962 at the Free University of Berlin (West). He was a fellow of the Airlift Memorial Foundation and received much of his microbiological training at the Pasteur Institute of Paris.

After a period of research and teaching at the Free University, he left Germany in 1967 and went to Canada, where he started to investigate bacteriophage morphology. During his career, he has done research on enterobacteria, airborne fungi, human hepatitis B virus, and baculoviruses; however, bacteriophages have always been the center of his interest. He teaches virology and mycology and has a strong interest in audiovisual teaching aids. In 1982, he founded the Félix d'Hérelle Reference Center for Bacterial Viruses, which is essentially a culture collection aimed at the preservation of type phages. Dr. Ackermann is the author of about 130 scientific papers and book chapters and of one book, entitled *Viruses of Prokaryotes* (CRC Press, 1987). He has been a member of the International Committee on Taxonomy of Viruses (ICTV) since 1971 and several times was chairman or vice-chairman of its Bacterial Virus Subcommittee and finally vice-president of the ICTV (1984 to 1990). He is presently a member of the ICTV Executive Committee.

Laurent Berthiaume, Ph.D., is a professor at the Virology Research Center of the Armand-Frappier Institute of the University of Quebec, located at Ville-de-Laval, a sister city of Montreal. He was born in Montreal in 1941 and obtained his Ph.D. degree in microbiology-immunology in 1972 at the University of Montreal. He was trained in electron microscopy at the University of Toronto School of Hygiene and for many years was in charge of the electron microscopic laboratory of this institute. He is presently coordinator of graduate studies.

During his activities as an electron microscopist, Dr. Berthiaume developed a strong interest in virus morphology and the rapid diagnosis of viral infections. He lectures on virus structure and taxonomy. His most recent work has centered on viruses infecting fish, especially on the antigenic and genetic diversity of aquabirnaviruses. He is the author of about 75 scientific papers or book chapters and 130 communications at congresses. He has been an ICTV member since 1975 and a member of the ICTV Executive Committee since 1989.

ACKNOWLEDGMENTS

This book would have been impossible to produce without the collaboration of several major publishers. First and foremost, the combined branches of Academic Press provided no less than 80 illustrations. This large number reflects the outstanding role of Academic Press as the publisher of such important periodicals as *Advances in Virus Research*, *Journal of Molecular Biology*, *Journal of Ultrastructure Research*, and *Virology*. Elsevier Science Publishers, Amsterdam, provided 22 diagrams. Most of them were from a single book entitled *Animal Virus Structure* (Reference 10), whose editors had arranged for illustrations of the highest quality. The American Society for Microbiology, publisher of the *Journal of Virology* and a series of important virological books, provided 17 diagrams. Thirteen more drawings, mostly from the *Journal of General Virology*, were obtained from the Society for General Microbiology (U.K.).

We wish to thank the following authors for permission to use their figures in this book: T. Alatossava, E. Arnold, M. Aymard, D.H. Bamford, W. Baschong, A.J.D. Bellett, P. Bentvelzen, M. Bergoin, D.H.L. Bishop, L.W. Black, H.E. Blum, D.P. Bolognesi, D.E. Bradley, D.T. Brown, J.C. Brown, S. Casjens, D. Cavanagh, H. Champsaur, Zh. Chen, R.W. Compans, P. Cornuet, R.A. Crowther, F. Darcy, F.W. Doane, F.A. Eiserling, J. Esparza, B.A. Federici, F. Fenner, J.T. Finch, H. Frank, C. Friend Norton, H. Garoff, H. Gelderblom, A.J. Gibbs, H.S. Ginsberg, M. Girard, R. Goldbach, C.F. Gonzalez, S.C. Harrison, T. Hatta, J. Hay, K. Hayashi, K.-H. Heermann, F.X. Heinz, R.W. Hendrix, K.V. Holmes, R.W. Horne, M.C. Horzinek, J.M. Huraux, I. Katsura, E. Kellenberger, D.W. Kingsbury, Ya. Kishko, E. Kurstak, G. Langenberg, L. Liljas, W.F. Liljemark, M. Longson, C.R. Madeley, R.E.F. Matthews, H. Matsuo-Kato, P. Metcalf, R.G. Milne, L. Mindich, L. Montagnier, G. Mosig, G. Müller, H.K. Narang, K. Nazerian, R. Neave, M.V. Nermut, A.R. Neurath, W.W. Newcomb, F. Nienhaus, J.S. Oxford, E.L. Palmer, W. Paranchych, P. Payment, R.F. Pettersson, S. Riva, G.F. Rohrmann, B. Roizman, M.G. Rossmann, R. Rott, R.R. Rueckert, W.C. Russell, H.L. Sänger, N.H. Sarkar, H. Scheid, L.A. Schiff, L.M. Stannard, W.A. Stevens, S. Sutherland, A.S. Tikhonenko, R. Vainionpää, K. Valegård, P. Vilaginès, J.D. Watson, W.H. Wunner, and W. Zillig.

F.A. Eiserling, H. Gelderblom, M.C. Horzinek, M.V. Nermut, P. Payment, G.F. Rohrmann, L. Thibodeau, and W.H. Wunner provided original or updated versions of already published diagrams or pointed to drawings that had escaped our attention. Jeannine Gauthier of Laval University and Zahira Chouchane, now in Sétif, Algeria, carried out an extensive bibliographical research. Ginette Larose of the Armand-Frappier Institute and Michel Côté, Rexfor, Quebec, prepared drawings. Gilles Mongrain, Laval University, and Mrs. Gauthier photographed and printed the diagrams shown. Céline Cinq and Andrée Jobin, Laval University, typed the manuscript with great care. Dr. Morris Goldner, also from this university, and Dr. Jacqueline Lecomte of the Armand-Frappier Institute provided valuable advice on presentation and editorial matters.

CONTENTS

Chapter 1
INTRODUCTION

Virology makes extensive use of diagrams for understanding and communicating information. Diagrams are derived from electron micrographs, mainly from negative staining and sectioning of viruses, and a variety of physicochemical, biological, and genetic techniques. The literature includes atlases of vertebrate, invertebrate, plant, and bacterial viruses. Some of them are concerned with the identification of human pathogens and several have contributed illustrations to this book.[1-5] All atlases are based on electron micrographs. There is no atlas of virus diagrams in existence; surprisingly, the many virus drawings in the literature have never been collected in a single volume despite their high didactic value.

Micrographs and diagrams are fundamentally different. Micrographs are documents and illustrations; diagrams are concepts, summaries, and illustrations. As a rule, a micrograph shows individual virus particles observed at a specific moment. It illustrates morphology alone and rarely shows all observable features of a virus.

A virus diagram is of a different nature. Whether naturalistic or abstract, it always sums up many morphological observations and often combines them with physicochemical data, such as genomic maps, and the location of constitutive proteins. Only a diagram can show, for example, both surface and interior of a virus or an exploded view. A diagram has always a higher information content than a micrograph. A diagram is also a concept that depends on the state of knowledge of the moment and the sense of beauty or abstraction of its author. It is fascinating to compare drawings of the same virus by different people and to see how concepts differ from one author to another. Some virus diagrams in the literature are definitely ugly; others, made by dedicated electron microscopists and gifted illustrators, are genuine works of art. The dependency of diagrams on possibly inaccurate data and personal interpretation makes them prone to error. On the other hand, some virus diagrams from the years 1950 to 1960 are of considerable historical interest, in particular, early drawings of tailed phages, poxviruses, and the tobacco mosaic virus indicate a surprising prescience and understanding of virus structure despite limited technical means. Showing them is a tribute to the scientists of these days. Diagrams are also interesting because, much more than micrographs, they are time capsules; indeed, the comparison of early and recent diagrams is a measure of the progress (or standstill) or virology itself. Finally, diagrams are prime teaching aids. It is hoped that this selection of drawings will stimulate the creation of more and better diagrams.

Diagrams were selected according to the following, sometimes conflicting, criteria: scientific value and information content, clarity and didactic potential, esthetic appeal, historic interest, diversity, and originality of concept. If several versions of the same diagram had been published by different people, we usually selected the original version. However, if these versions were from the same author, we preferred the last, updated diagram. Some virus types were illustrated in virtually every textbook, the greatest favorites being herpesviruses and retroviruses. On the other hand, the literature included no detailed diagrams of fungal and algal viruses and filamentous plant viruses were under-represented.

The surveyed literature included about 300 books or monographs and 50 periodicals from the virological literature in the English, French, and German languages. Parts of the Polish, Russian, and Spanish literature were surveyed, too. In some instances, when no diagram of a particular virus could be found in the literature or the material was unsatisfactory, we included our own drawings. Some of these were prepared after models.

This book contains 270 drawings of complete viruses or virus parts. It is concerned with the morphological aspects of comparative virology, but is no textbook or treatise of virus taxonomy. The principles of virus structure are not discussed here. They are explained in detail in many textbooks, and particularly in references.[6-11] Further, this book does not include virus models, molecular models of virus proteins, electron density maps, and diagrams derived from rotational analysis and image filtering. These subjects are outside the essentially comparative scope of this book.

A few interesting and artistically satisfactory diagrams were omitted because of prohibitive fees, for example, for the right to reproduce a single diagram. If generalized, such fees would have raised the starting costs for producing this volume to over $100,000. We included here indeed a few expensive diagrams, for example, from *Scientific American*, but we feel that the perception of hefty fees is detrimental to scientific publications and essentially destructive. The present copyright regulations should be revised.

The number of illustrations depends on the virus group and reflects several factors: current research interests, the intrinsic complexity of a virus and difficulties in understanding its structure, and, quite simply, the amount of money and time invested into its study. It is thus not surprising that retroviruses range first with over 20 diagrams, followed by reoviruses and poxviruses. Diagrams are usually reproduced as found in the literature. French and German legends that are part of a figure are reproduced without modification. It is hoped that they will be understood from the context.

Diagrams are grouped by virus morphology and not nucleic acid characteristics. Viruses consist basically of a capsid and nucleic acid. The capsid is of cubic or helical symmetry and may be surrounded by a lipid-containing envelope that is derived from cell membranes (e.g., nuclear, cytoplasmic) or synthetized *de novo*. Capsids are icosahedra or derivatives thereof. Tailed phages are combinations of "cubic" capsids and helical tails. There is growing evidence that (1) numerous plant and animal RNA viruses of different morphology and (2) tailed phages are phylogenetically related. Consequently, virus diagrams are grouped into three chapters:

1. OVERVIEWS

Diagrams illustrate host-related or biological groups such as vertebrate or bacterial viruses, are composite, and show few details. Similar overviews are published by the International Committee on Taxonomy of Viruses (ICTV).[12]

2. VIRUSES WITH CUBIC OR HELICAL SYMMETRY

This large chapter comprises more or less detailed diagrams of isometric and filamentous viruses of vertebrates, invertebrates, plants, and bacteria. Diagrams are grouped by virus family or grouped in alphabetical order; however, there are a few exceptions to save space (*Filoviridae, Plasmaviridae*). The last section also includes drawings of defective virus, virus-like particles, and viroids.

3. VIRUSES WITH BINARY SYMMETRY

These viruses, characterized by a combination of constituents with cubic and helical symmetry, are commonly known as tailed bacteriophages or "phages" and have been illustrated many times since the advent of electron microscopy; indeed, coliphages are the very first viruses represented by a diagram (Figure 207). Not surprisingly, T-even phages are illustrated many times. Tailed phages are extremely diversified and include a

large number of species. The literature comprises a considerable number of classification schemes for phages of specific bacteria, e.g., pseudomonads and vibrios. A sample of these diagrams is shown here to illustrate the variety of phage species that can be found in a few important bacterial groups such as bacilli and enterobacteria (Figures 257 to 263). The chapter concludes with representations of defective phages and morphological aberrations.

Each section or subdivision is preceded by a "box" and a short description of the virus group illustrated. The box contains basic properties such as the nature of the nucleic acid, presence or absence of an envelope, and the symmetry of the capsid. The description focuses on size and taxonomic structure of virus groups, host range, pathogenicity, and such morphological features that are necessary for understanding the diagrams. The presentation of diagrams depends on the material available. Some sections start with a particularly detailed diagram and others with the taxonomical structure of a virus family or with historical drawings. An attempt is made to group diagrams by year of publication and content, starting with views of complete viruses.

Legends are short as not to distract from the diagrams. The original legends varied greatly in length and content. Some extended over whole pages and others just read "phage T2" or "schematic representation." Others, taken out of context, were difficult to understand. Legends were thus generally paraphrased and shortened. Some were completed, for example, by specifying the taxonomic position of certain plant viruses. The information content of the original legends was always preserved, even if we did not agree with it.

A final cautionary remark is needed. There are several conventions for representing data and features, making virus diagrams something of a modern pictograph. For example, capsids with cubic symmetry are frequently shown as hexagons with regular arrays of capsomers. This is acceptable in adenoviruses which have sharp outlines and clearly visible capsomers, but is more problematic in papovaviruses or picornaviruses which usually appear in the electron microscope as small round particles without surface structures. Similarly, vertebrate iridoviruses, due to the presence of lipids in the capsid, and heads of tailed phages often appear swollen or rounded in the electron microscope. Representations of hexagonal capsids, though probably rendering the true shape of these viruses, constitute thus a simplification. Other conventions are the representation of envelopes and viral genomes as simple lines or of genome-associated proteins as black dots.

Chapter 2

A SUMMARY OF VIRUS CLASSIFICATION

The classification of viruses, first limited to the efforts of an individual virologist, became a worldwide concerted enterprise in 1963 with the establishment of the International Committee on Nomenclature of Viruses (ICNV). The committee became permanent in 1966 at the XIth International Microbiology Congress in Moscow and in 1973 was renamed the International Committee on Taxonomy of Viruses (ICTV).[13]

The ICTV presently consists of 8 subcommittees, 45 study groups, and over 400 participating virologists. It has subcommittees for viruses of vertebrates, invertebrates, plants, bacteria, and eukaryotic protists (fungi, alga, protozoa). In addition, there are subcommittees for coordination (for viruses infecting more than one kind of hosts) and the creation of a viral database. Study groups are formed as the need arises. Taxonomic and nomenclature proposals are typically made in study groups, voted on by the respective subcommittees, scrutinized by the ICTV Executive Committee, and ratified (or not) by the full ICTV when it convenes every 3 years at an International Congress of Virology. Only then do they become official.

The ICTV is concerned with the construction of a coherent and universal system of virus classification and nomenclature. The ICTV collects data and proposals and issues periodically a report in principle after each International Congress of Virology. The report describes virus families and genera, but not species, and contains updates of existing taxa and descriptions of new groups. The Sixth Report has just appeared. As a novelty, it contains numerous micrographs and genomic maps.[12]

In principle, viruses are classified with the help of all available criteria. However, some criteria are of particular value, namely the presence of DNA or RNA, capsid symmetry, and the presence or absence of an envelope. These criteria were introduced in 1962 by Lwoff et al.[14] together with a hierarchical system of viruses. The system did not survive, but most of its criteria stood the test of time. More high-level criteria have been added since. They relate to the genome (strandedness, message-sense [−] or anti-message [+] strands, number of segments) and replication (presence of a DNA step or RNA polymerase). In essence, viruses are classified by virus-dependent, morphological, physicochemical, and physiological criteria. Host-related data such as symptoms of disease are of limited importance. Genome structure and amino acid alignments are becoming increasingly important as data accumulate. The ultimate goal is a phylogenetic system of viruses. Viruses are essentially classified by the following criteria:

1. Genome: DNA, RNA, number of strands, linear, circular, superhelical, sense (+ or −), size, number and size of segments, G+C content, nucleotide sequence, presence or absence of 5′-terminal caps, terminal proteins, or poly(A) tracts.
2. Morphology: Size, shape, presence, or absence of an envelope and peplomers, capsid size, and structure.
3. Physicochemical properties: Mass, buoyant density, sedimentation velocity, and stability (pH, heat, solvents, detergents).
4. Proteins: Content, number, size, and function of structural and nonstructural proteins, amino acid sequence, glycosylation.
5. Lipids and carbohydrates: Content and nature.
6. Genome organization and replication: Gene number and genomic map, characteristics of transcription and translation, post-translational control, site of viral assembly, type of release.
7. Antigenic properties.
8. Biologic properties: Host range, mode of transmission, geographic distribution, cell and tissue tropisms, pathogenicity and pathology.

The present edifice of virus classification includes a single order, the *Mononegavirales*, characterized by the presence of a single molecule of negative-sense ssRNA. The order includes the families *Filoviridae, Paramyxoviridae,* and *Rhabdoviridae.* Tailed bacteriophages are likely to become another order. The system includes 50 families and over 160 genera. The suffix *-viridae* is the ending for family names; genus names end with the suffix *-virus.* A considerable effort is made to devise names that express viral characteristics. A number of "floating" genera are not yet classified into families. They are particularly frequent in plant viruses. The *Poxviridae* family has 11 genera, but other virus families are monogeneric and some even consist of a single member. The ICTV has also adopted the polythetic species concept, meaning that a species is defined by a set of properties, all of which are not necessarily present in all members. A virus species is defined as a polythetic class of viruses that constitutes a replicating lineage and occupies a particular ecological niche.[13]

The state of virus taxonomy is summarized in Table 1. Virus taxa are more or less listed in the order of the Sixth ICTV Report.[12] While most viruses have cubic or helical nucleocapsids, a few viruses (Fuselloviridae, Plasmaviridae, Umbraviridae) apparently have an envelope only and no capsid, and cannot be attributed with certainty to any symmetry class.

Table 1
VIRUS FAMILIES AND "FLOATING" GENERA

Genome	Envelope	Capsid Symmetry	Family	Floating Genera	Number of Genera	Host[a]
dsDNA	Yes	Cubic	Hepadnaviridae		2	V
			Herpesviridae		7	V, I
		Helical	Baculoviridae		2	I
			Lipothrixviridae		1	B
			Polydnaviridae		2	I
			Poxviridae		11	V, I
		None	Fuselloviridae		1	B
			Plasmaviridae		1	B
	No	Binary	Myoviridae		1	B
			Siphoviridae		1	B
			Podoviridae		1	B
		Cubic	Adenoviridae		2	V
			Corticoviridae		1	B
			Iridoviridae		5	V, I
			Papovaviridae		2	V
			Phycodnaviridae		1	A
			Tectiviridae		1	B
			—	ASFV[b]	1	V
			—	Badnavirus	1	P
			—	Caulimovirus	1	P
			—	Rhizidiovirus	1	P
ssDNA	No	Cubic	Circoviridae		1	P
			Geminiviridae		3	P
			Microviridae		4	B
			Parvoviridae		6	V, I
		Helical	Inoviridae		2	B
dsRNA	Yes	Cubic	Cystoviridae		1	B
		None	Hypoviridae		1	F
	No	Cubic	Birnaviridae		3	V, I
			Partitiviridae		4	F, P
			Reoviridae		9	V, I, P
			Totiviridae		3	F, Pr
		DNA Step in Replication				
ssRNA	Yes	Cubic	Retroviridae		7	V
		No DNA Step				
		Genome Negative-Sense, Nonsegmented				
		Helical	Mononegavirales			
			Filoviridae		1	V
			Paramyxoviridae		4	V
			Rhabdoviridae		5	V, I, P
			—	Deltavirus	1	V

Table 1
VIRUS FAMILIES AND "FLOATING" GENERA (CONTINUED)

Genome	Envelope	Capsid Symmetry	Family	Floating Genera	Number of Genera	Host[a]
			Genome Negative-Sense, Segmented			
		Helical	*Arenaviridae*		1	V
			Bunyaviridae		5	V, I, P
			Orthomyxoviridae		3	V, I
	No		—	*Tenuivirus*	1	P
			Genome Positive-Sense			
ssRNA	Yes	Helical	*Coronaviridae*		3?	V
		Cubic	*Flaviviridae*		3	V, I
			Togaviridae		2	V, I
	No	Cubic	*Astroviridae*		1	V
			Bromoviridae		4	P
			Caliciviridae		1	V
			Comoviridae		3	P
			Leviviridae		2	B
			Nodaviridae		1	I
			Picornaviridae		5	V
			Sequiviridae		1	P
			Tetraviridae		2	I
			Tombusviridae		2	P
			—	*Dianthovirus*	1	P
			—	*Enamovirus*	1	P
			—	*Idaeovirus*	1	P
			—	*Luteovirus*	1	P
			—	*Machlomovirus*	1	P
			—	*Marafivirus*	1	P
			—	*Necrovirus*	1	P
			—	*Sobemovirus*	1	P
			—	*Tymovirus*	1	P
			Rods			
		Helical	*Barnaviridae*		1	F
			—	*Furovirus*	1	P
			—	*Hordeivirus*	1	P
			—	*Tobamovirus*	1	P
			—	*Tobravirus*	1	P
			Filaments			
			Potyviridae		3	P
			—	*Capillovirus*	1	P
			—	*Carlavirus*	1	P
			—	*Closterovirus*	1	P
			—	*Potexvirus*	1	P
			—	*Trichovirus*	1	P
	None		—	*Umbravirus*	1	P

[a]A, algae; B, bacteria; F, fungi; I, invertebrates; P, plants; Pr, protozoa; V, vertebrates. [b]ASFV, African swine fever virus.

_____ *Chapter 3* _____

OVERVIEWS

DNA VIRUSES

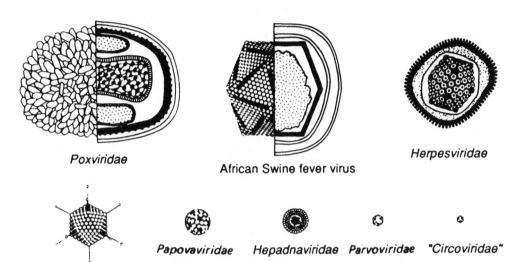

Poxviridae

African Swine fever virus

Herpesviridae

Adenoviridae

Papovaviridae   Hepadnaviridae   Parvoviridae   "Circoviridae"

RNA VIRUSES

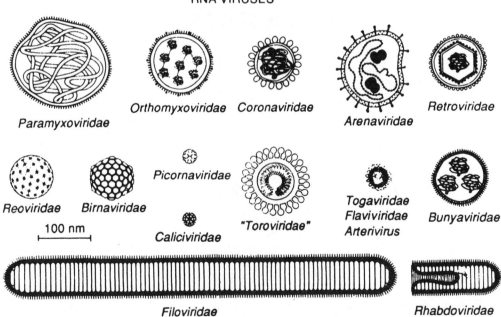

Paramyxoviridae

Orthomyxoviridae   Coronaviridae

Arenaviridae

Retroviridae

Reoviridae   Birnaviridae

Picornaviridae

Caliciviridae

"Toroviridae"

100 nm

Togaviridae
Flaviviridae
Arterivirus

Bunyaviridae

Filoviridae

Rhabdoviridae

FIGURE 1

Size and shape of vertebrate viruses. Scale drawings illustrating surface views, cross-sections, or a combination of both. (From Fenner, F.J., Gibbs, E.P.J., Murphy, F.A., Rott, R., Studdert, M.J., and White, D.O., *Veterinary Virology*, 2nd ed., Academic Press, 1993, 21. With permission.)

RNA VIRUSES

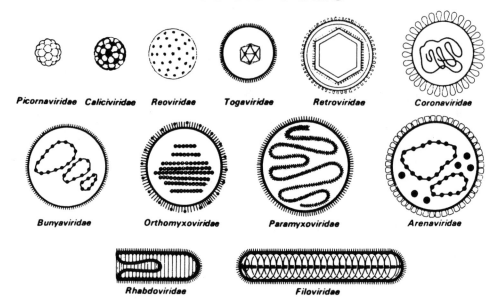

DNA VIRUSES

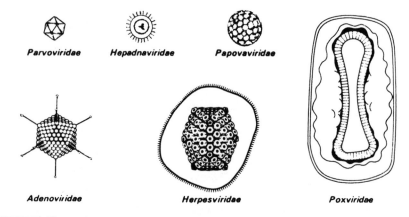

FIGURE 2
Approximate size and shape of vertebrate viruses; surface views and cross-sections. The filovirus is not to scale. (From Palmer, E.L. and Martin, M.L., *Electron Microscopy in Viral Diagnosis*, 1988, 2, CRC Press, Boca Raton, FL.)

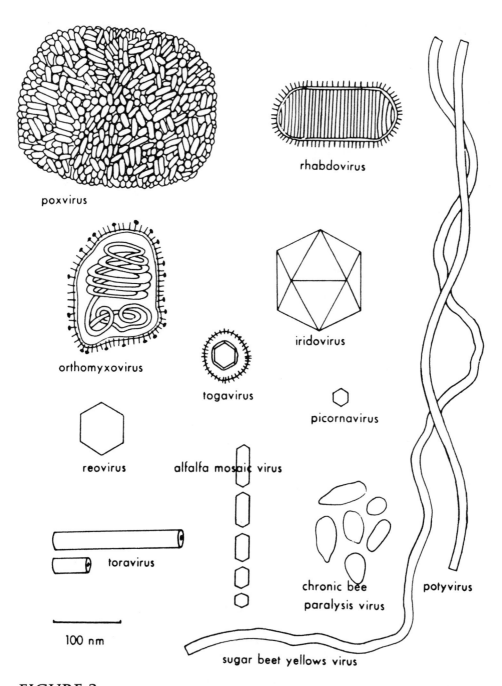

FIGURE 3

Approximate size and shape of viruses associated with invertebrates. The poxvirus diagram actually represents a vertebrate virus (author's note). (From Bellett, A.J.D., Fenner, F., and Gibbs, A.J., *Frontiers of Biology*, Vol. 3, Gibbs, A.J., Ed., Elsevier North-Holland, Amsterdam, 1973, 41. With permission.)

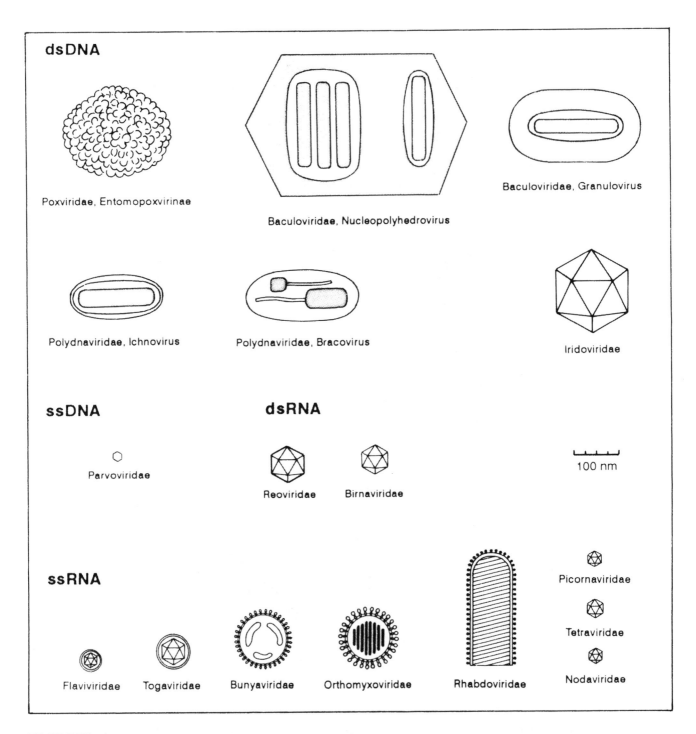

FIGURE 4

Size and shape of invertebrate viruses. (By H.-W. Ackermann and M. Côté, modified from Reference 18.)

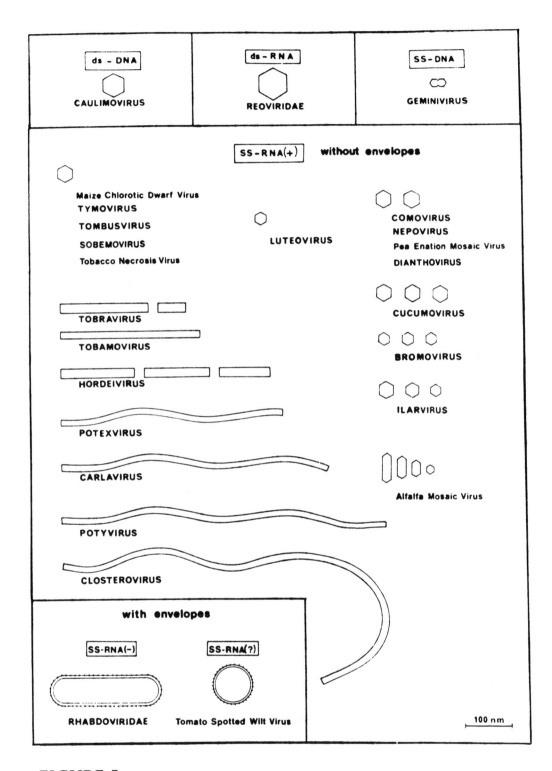

FIGURE 5

Size and shape of plant viruses classified by type of nucleic acid, representing several multipartite viruses *(Cucumovirus, Tobravirus,* others). The tomato spotted wilt virus is now classified as a bunyavirus. (From Francki, R.I.B., Milne, R.G., and Hatta, T., *Atlas of Plant Viruses,* Vol. 1, 1985, 9, CRC Press, Boca Raton, FL.)

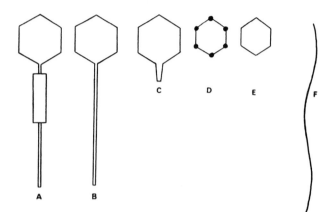

FIGURE 6

Basic morphological types of bacteriophages. This diagram from 1965 represents the first attempt at phage classification and is still widely cited. Present classification: A, *Myoviridae;* B, *Siphoviridae;* C, *Podoviridae;* D, *Microviridae;* E, *Leviviridae;* F, *Inoviridae (Inovirus* genus). (From Bradley, D.E., *J. R. Microsc. Soc.,* 84, 275, 1965. With permission.)

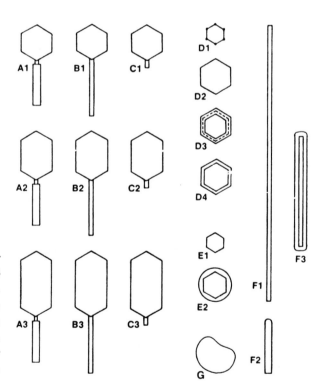

FIGURE 7

A more recent classification scheme (1987) based on Figure 6, showing new groups and subdivision of tailed phages by head shape: A1–3, *Myoviridae;* B1–3, *Siphoviridae;* C1–3, *Podoviridae;* D1, *Microviridae;* D3, *Corticoviridae;* D4, *Tectiviridae;* E1, *Leviviridae;* E2, *Cystoviridae;* F1, *Inoviridae, Inovirus* genus; F2, *Inoviridae, Plectrovirus* genus; F3, *Lipothrixviridae;* G, *Plasmaviridae;* D2 is for unclassified isometric DNA phages. (From Ackermann, H.-W. and DuBow, M.S., *Viruses of Prokaryotes,* Vol. 1, 1987, 16.)

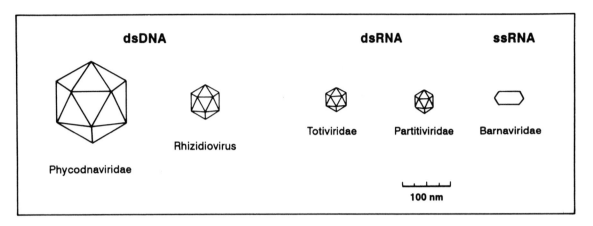

FIGURE 8
Fungal, algal, and protozoal viruses. (By H.-W. Ackermann and M. Côté, modified from Reference 18.)

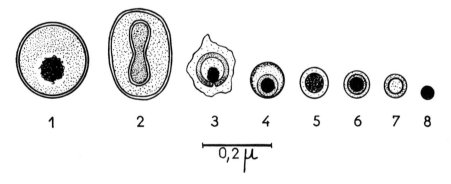

FIGURE 9
Tumor viruses. 1. and 2. Immature and mature poxviruses (Shope fibroma and molluscum contagiosum). 3. Lucké frog virus. 4. Bittner virus. 5. Virus associated with mouse leukemia. 6. Avian tumor and leukemia viruses. 7. A-type particle found in various mouse tumors. 8. Polyoma-papilloma type. (From Bernhard, W., *Cancer Res.*, 20, 712, 1960. With permission.)

Chapter 4

VIRUSES WITH CUBIC OR HELICAL SYMMETRY

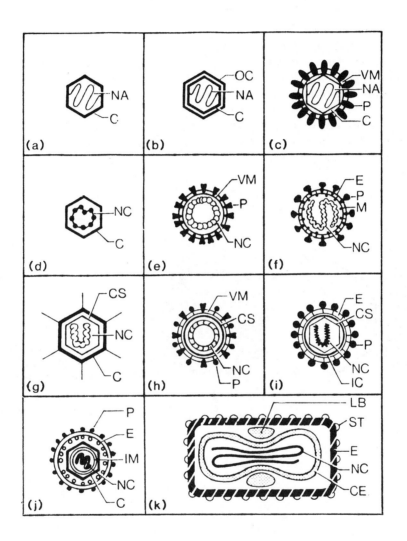

FIGURE 10

Architecture of animal viruses and attempt at a unifying morphological nomenclature. (a) Naked icosahedral nucleocapsid. (b) Encapsidated icosahedral nucleocapsid. (c) Enveloped icosahedral nucleocapsid. (d) Encapsidated "nucleosomes." (e) and (f) Two types of enveloped helical nucleocapsids. (g) Encapsidated core consisting of "helical" nucleocapsid and core shell. (h) Enveloped icosahedral core (helical nucleocapsid and core shell). (i) Enveloped icosahedral core with helical (?) nucleocapsid. (j) Enveloped icosahedral core with linear (?) nucleocapsid. (k) Enveloped complex core with linear nucleocapsid, core envelope, and lateral bodies. C, Capsid; CE, core envelope; CS, core shell; E, envelope; IC, inner coat protein; IM, inner lipid membrane; LB, lateral body; M, membrane protein; NA, nucleic acid; OC, outer capsid; P, peplomers or surface projections; ST, surface tubule; VM, virus membrane. (From Nermut, M.V., *Animal Virus Structure,* Nermut, M.V. and Steven, A.C., Eds., Elsevier, Amsterdam, 1987, 3. With permission.)

dsDNA, linear
Cubic, naked
Vertebrates

I. ADENOVIRIDAE

These were first isolated from human adenoid tissue including mammalian and avian viruses with numerous serotypes (genera *Mastadenovirus* and *Aviadenovirus*). Particles are nonenveloped icosahedra of 70 to 90 nm in diameter. The capsid consists of 252 capsomers (T=25), with 240 hexons (6 subunits) located on sides and edges of the particle and 12 apical pentons (5 subunits) provided with one or two thin fibers. The core seems to consist of a string of nucleosome-like particles.

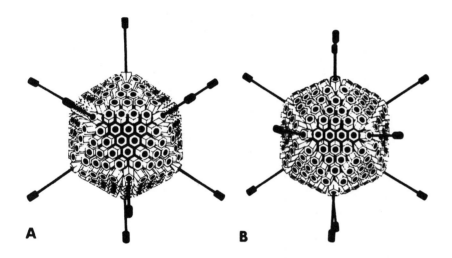

FIGURE 11
Adenoviruses showing hexons, pentons, and fibers, viewed along axes of three-fold (A) and twofold (B) symmetry. (From Horne, R.W., Pasquali Ronchetti, I., and Hobart, J.M., *J. Ultrastruct. Res.*, 51, 233, 1975. With permission.)

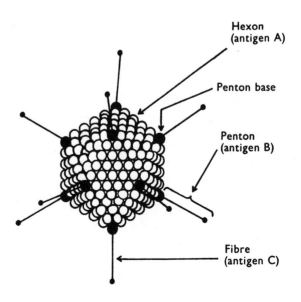

FIGURE 12
Location of adenovirus antigens. (From Russell, W.C., Hayashi, K., Sanderson, P.J., and Pereira, H.G., *J. Gen. Virol.*, 1, 495, 1967. With permission.)

ADENOVIRIDAE

Cross-sections showing location of DNA and antigens

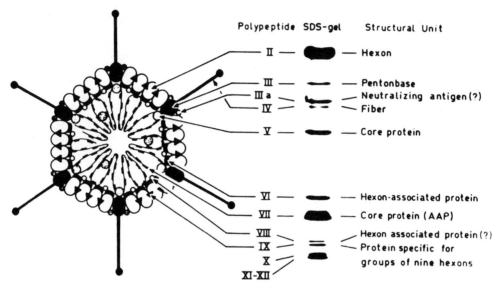

FIGURE 13
(From Everitt, E., Lutter, L., and Philipson, L., *Virology*, 67, 197, 1975. With permission.)

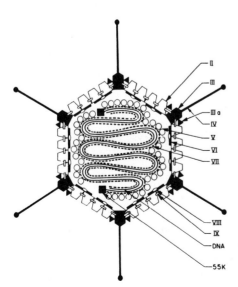

FIGURE 14
(From Ginsberg, H.S., *Comprehensive Virology*, Vol. 13, Fraenkel-Conrat, H. and Wagner, R.R., Eds., Plenum Press, New York, 1979, 49. With permission.)

ADENOVIRIDAE

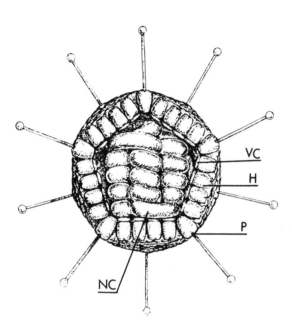

FIGURE 15

Interior view. H, hexon; NC, nucleocapsid; P, penton; VC, virus core. (From Nermut, M.V., *Arch. Virol.*, 64, 175, 1980. With permission.)

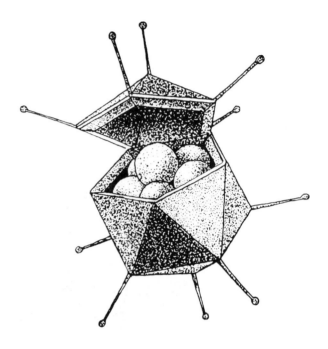

FIGURE 16

Alternative view of adenovirus interior with representation of the virus-protein complex by 12 globular bodies. (From Nermut, M.V., *Animal Virus Structure*, Nermut, M.V. and Steven, A.C., Eds., Elsevier, Amsterdam, 1987, 373. With permission.)

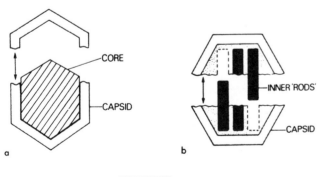

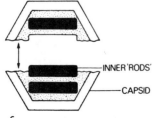

FIGURE 17

Viruses freeze-fractured between capsid and core (a) or with "inner rods" in vertical or horizontal orientation (b and c). (From Nermut, M.V., *Arch. Virol.*, 57, 323, 1978. With permission.)

ADENOVIRIDAE

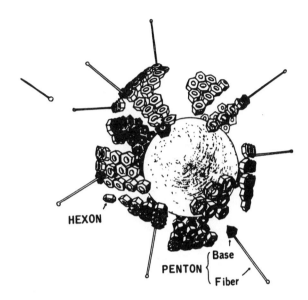

HEXON

PENTON {
Base

Fiber

FIGURE 18

Partially disrupted type 5 adenovirus particle. (From Ginsberg, H.S. and Young, C.S.H., *Comprehensive Virology*, Vol. 9, Fraenkel-Conrat, H. and Wagner, R.R., Eds., Plenum Press, New York, 1977, 27. With permission.)

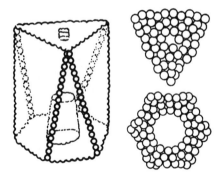

FIGURE 19

Hexon model derived from a 6-Å electron density map, consisting of 1200 spheres arranged in 18 layers. (From Berger, J., Burnett, R.M., Franklin, R.M. and Grütter, M., *Biochim. Biophys. Acta*, 535, 233, 1978. With permission.)

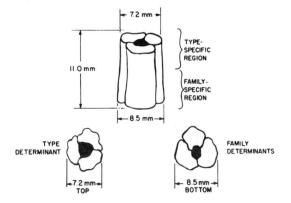

FIGURE 20

Hexon model showing family- and type-specific antigens. (From Ginsberg, H.S., *Comprehensive Virology*, Vol. 13, Fraenkel-Conrat, H. and Wagner, R.R., Eds., Plenum Press, New York, 1979, 409. With permission.)

ssRNA, 2 segments
Helical, enveloped
Vertebrates

II. ARENAVIRIDAE

This is a family with a single genus of essentially rodent-pathogenic viruses and includes the agent of Lassa fever. Particles are spherical or pleomorphic and usually 110 to 130 nm in size. The envelope has club-shaped surface projections and contains two closed circles of nucleoprotein and a variable number of ribosomes.

Arenavirus cross-sections

G1 and G2, surface glycoproteins derived by cleavage from a common precursor; L, putative RNA polymerase; N, N protein; NP, nucleoprotein; vRNA, viral RNA.

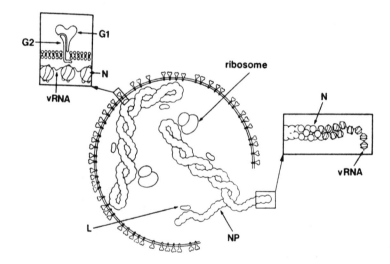

FIGURE 21
(From Young, P.R., *Animal Virus Structure*, Nermut, M.V. and Steven, A.C., Eds., Elsevier, Amsterdam, 1987, 195. With permission.)

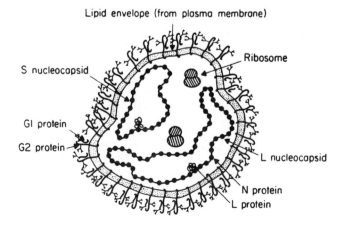

FIGURE 22
(From Bishop, D.H.L., *Immunobiology and Pathogenesis of Persistent Virus Infections*, Lopez, C., Ed., American Society for Microbiology, Washington, DC, 1988, 79. With permission.)

dsDNA, circular
Helical, enveloped
Invertebrates

III. BACULOVIRIDAE

Baculoviruses infect insects and crustacea (shrimps). Viruses are generally occluded in virus-specified protein crystals and consist of an envelope containing one or more cylindrical nucleocapsids of 220 to 400 × 50 nm. The family has two genera:

1. *Polyhedrovirus:* Occlusion bodies are polyhedra of 1 to 4 μm in size and contain 20 to 200 viruses with a single or multiple (up to 32) nucleocapsids per envelope.
2. *Granulovirus:* Occlusion bodies are "granules" of about 0.2 to 0.4 μm and contain one virus with a single nucleocapsid.

Occluded baculoviruses

FIGURE 23
Viruses drawn after electron micrographs: multiple-enveloped (A) and single-enveloped (B) nuclear polyhedrosis viruses; granulosis virus (C). (By H.-W. Ackermann.)

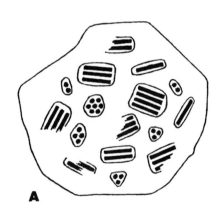
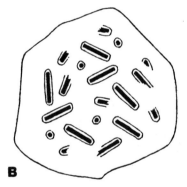
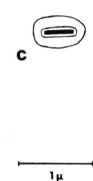

A

B

C

1 μ

Polyeder

einzeln eingeschlossenes
Virion

Granulum

Polysacharidlamelle
Virion
unit-Membran
Nukleokapsid
Proteinlamelle

Polyhedrin

multipel eingeschlossene
Virionen

Granulin

FIGURE 24
A more schematic view. (With the kind permission of Blackwell Wissenschafts-Verlag GmbH. Taken from Horzinek, C., *Kompendium der allgemeinen Virologie,* 2nd ed., Paul Parey, Berlin, 1985, 51.)

BACULOVIRIDAE

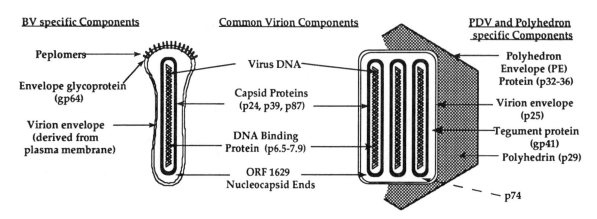

Budded Virus (BV)   **Polyhedron Derived Virus (PDV)**

BV specific Components   Common Virion Components   PDV and Polyhedron specific Components

Peplomers

Envelope glycoprotein (gp64)

Virion envelope (derived from plasma membrane)

Virus DNA

Capsid Proteins (p24, p39, p87)

DNA Binding Protein (p6.5-7.9)

ORF 1629 Nucleocapsid Ends

Polyhedron Envelope (PE) Protein (p32-36)

Virion envelope (p25)

Tegument protein (gp41)

Polyhedrin (p29)

p74

FIGURE 25
Structural components of budding (BV) and polyhedron-contained (PDV) baculovirus phenotypes. Envelopes are acquired from the plasma membrane or assembled in the nucleus. The location of protein p74 has not been determined. (From Rohrmann, G.F., *J. Gen. Virol.*, 73, 749, 1992. With permission.)

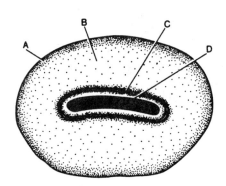

FIGURE 26
Section of intact granulosis virus. A. Cortical layer of granulin. B. Granulin. C. Envelope with dense material deposited on both surfaces. D. Nucleocapsid. (From David, W.A.L., *Adv. Virus Res.*, 22, 111, 1978. With permission.)

FIGURE 27
Bundle of nuclear polyhedrosis virus capsids contained within the same envelope; combined length and cross-section. (From Kawamoto, F. and Asayama, T., *J. Invert. Pathol.*, 26, 47, 1975. With permission.)

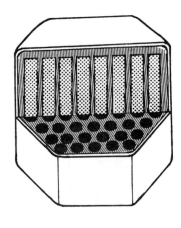

BACULOVIRIDAE

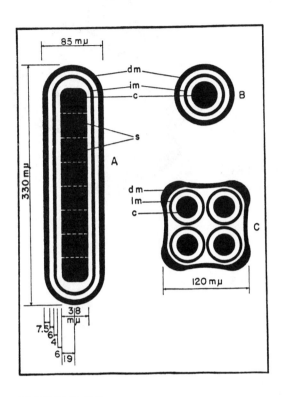

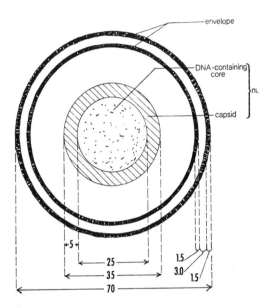

FIGURE 28

Nuclear polyhedrosis viruses in length and cross-sections. A. and B. Virus of *Bombyx mori*. C. Virus of *Laphygma frugiperda*: c, core; d, developmental membrane; im, intimate membrane. (From Bergold, G.H., *J. Insect Pathol.*, 5, 11, 1963. With permission.)

FIGURE 29

Cross-section of a single-enveloped nuclear polyhedrosis virus; dimensions in nanometers. (From Hughes, K.M., *J. Invert. Pathol.*, 19, 198, 1972. With permission.)

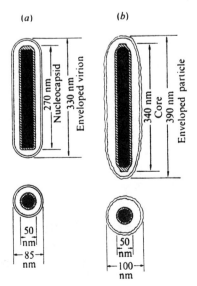

FIGURE 30

A baculovirus of *Bombyx mori* (a) compared to a baculovirus-like particle of the insect-pathogenic fungus, *Strongwellsea magna* (b). (From Federici, B.A. and Humber, R.A., *J. Gen. Virol.*, 35, 387, 1977. With permission.)

BACULOVIRIDAE

FIGURE 31
Six-armed nodal unit of polyhedron protein. (From Harrap, K.A., *Virology*, 50, 114, 1972. With permission.)

dsRNA, 2 segments,
Cubic, naked
Vertebrates, invertebrates

IV. BIRNAVIRIDAE

This is a small, widespread family of viruses infecting birds, fish, mollusks, and insects, characterized by the presence of two molecules of dsRNA (hence "bi-RNA"). Particles are naked icosahedra of 60 nm in diameter with 92 or 132 capsomers.

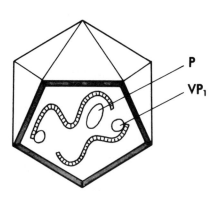

FIGURE 32
Opened capsid. Note the unequal length of RNA molecules; P, RNA polymerase. (By H.-W. Ackermann and M. Côté.)

ssRNA, 3 segments
Helical, enveloped
Vertebrates, invertebrates, plants

V. BUNYAVIRIDAE

This large family with five genera is named after Bunyamwera, a locality in Uganda, and includes many "arboviruses" and some plant pathogens (genus *Tospovirus)*. Particles are spherical to pleomorphic of 80 to 120 nm in size, and consist of an envelope and three circular nucleocapsids. The envelope contains glycoproteins that form spikes and, at least in some members, appear as cylindrical subunits distributed in a regular pattern (T=12?).

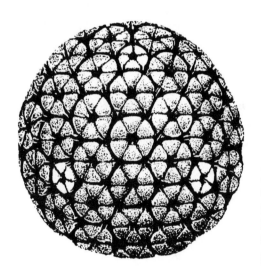

FIGURE 33
Clustered glycoproteins of Uukuniemi virus envelope arranged in a T=12 lattice. (From Pettersson, R.F. and von Bonsdorff, C.-H., *Animal Virus Structure,* Nermut, M.V. and Steven, A.C., Eds., Elsevier, Amsterdam, 1987, 147. With permission.)

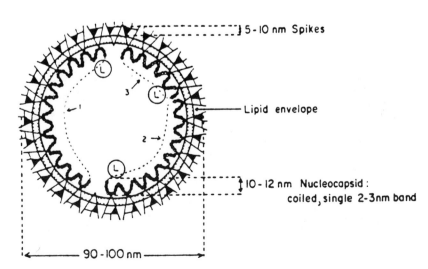

FIGURE 34
Sectioned virus showing surface spikes (G1 and G2 proteins) and three circular nucleocapsids. (From Bishop, D.H.L., *Virus Res.,* 31, 1, 1986. With permission.)

BUNYAVIRIDAE

Several views of tomato spotted wilt virus *(Tospovirus* **genus)**

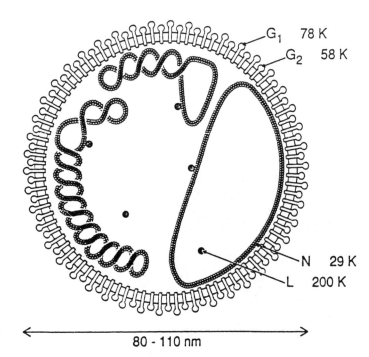

FIGURE 35
Section showing nucleocapsids. L, viral transcriptase (?); N, nucleocapsid protein. (From Goldbach, R. and De Haan, P., *Semin. Virol.*, 4, 381, 1993. With permission.)

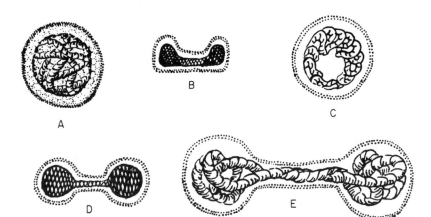

FIGURE 36
A and C. Particles seen from above. B. Section of particle sitting on electron microscopical grid. D and E. Dumbbell-shaped particles. (From Best, J.R., *Adv. Virus Res.*, 13, 65, 1968. With permission.)

ssRNA, + sense
Cubic, naked
Vertebrates

VI. CALICIVIRIDAE

These viruses occur in a wide variety of vertebrates and are frequently associated with gastroenteritis. Particles are naked icosahedra of 35 to 39 nm in diameter, characterized by 32 cup-shaped depressions (Latin *calix*, cup). The capsid (T=3) probably consists of 180 identical subunits arranged in 90 dimeric capsomers.

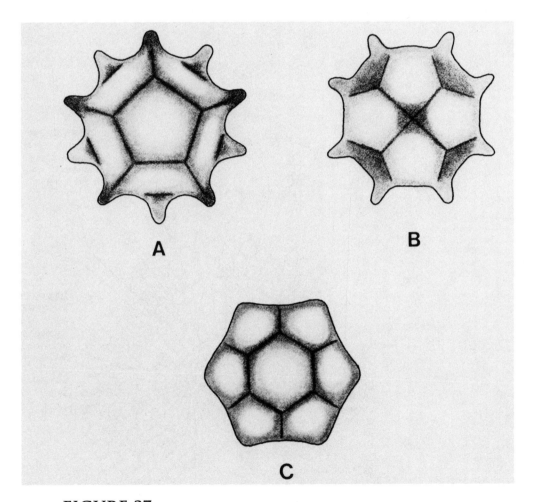

FIGURE 37
Calicivirus seen along fivefold (A), twofold (B), and threefold (C) axes of rotational symmetry. (By H.-W. Ackermann and M. Côté.)

ssRNA, + sense
Helical, enveloped
Vertebrates

VII. CORONAVIRIDAE

These viruses of mammals and birds are often associated with respiratory or enteric disease. Particles are spherical to pleomorphic and usually 80 to 120 nm in diameter. Members of the genus *Coronavirus* are characterized by a "crown" of large, 20-nm long surface projections. The genus *Torovirus* includes disk-, kidney-, or rod-shaped particles with tubular nucleocapsids. The genus *Arterivirus* (equine arteritis virus) was formerly a member of the *Togaviridae* family and has been included into the *Coronaviridae* on the basis of genome organization and strategy of replication. *Arterivirus* particles are about 60 nm in size.

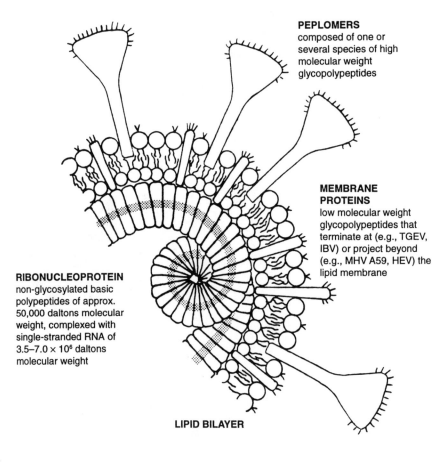

PEPLOMERS
composed of one or several species of high molecular weight glycopolypeptides

MEMBRANE PROTEINS
low molecular weight glycopolypeptides that terminate at (e.g., TGEV, IBV) or project beyond (e.g., MHV A59, HEV) the lipid membrane

RIBONUCLEOPROTEIN
non-glycosylated basic polypeptides of approx. 50,000 daltons molecular weight, complexed with single-stranded RNA of 3.5–7.0 × 10⁶ daltons molecular weight

LIPID BILAYER

25 nm

FIGURE 38
Molecular organization of coronaviruses. (From Garwes, D.J., *INSERM*, 90, 141, 1979. With permission.)

CORONAVIRIDAE

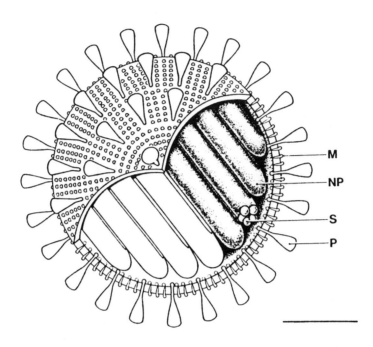

FIGURE 39

Scale model (bar = 50 nm). M, transmembrane protein; NP, nucleocapsid protein; P, surface projections; S, subunit of NC protein. (From Macnaughton, M.R. and Davies, H.A., *Animal Virus Structure,* Nermut, M.V. and Steven, A.C., Eds., Elsevier, Amsterdam, 1987, 173. With permission.)

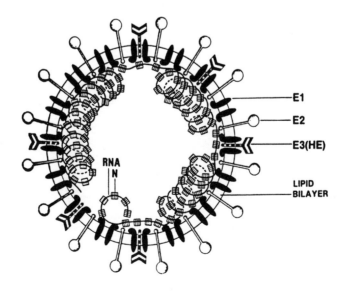

FIGURE 40

Section of particle. The nucleocapsid consists of RNA and phosphorylated glycoprotein N. The envelope consists of a lipid bilayer, derived from host viral membranes, and viral glycoproteins (E1, E2, E3 or HE). (From Holmes, K.V., *Virology,* 2nd ed., Fields, B.N. and Knipe, D.M., Eds., Raven Press, New York, 1990, 841. With permission.)

CORONAVIRIDAE

Berne virus, the *Torovirus* prototype

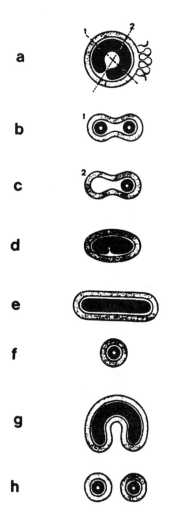

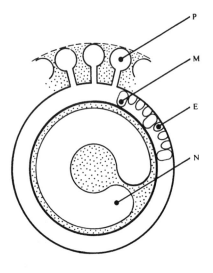

FIGURE 42

Cross-section indicating the location of proteins. E (envelope), M (matrix), N (nucleocapsid), and P (peplomers). (From Horzinek, M.C., Ederveen, J., Kaeffer, B., De Boer, D., and Weiss, M., *J. Gen. Virol.*, 67, 2475, 1986. With permission.)

FIGURE 41

Different types of sectioned particles. a. Virus with toroidal core and circular outline. b. Biconcave particle with twin cross-sections of core according to plane 1. c. Section of core in plane 2. d. Ellipsoidal particle with little resolution of content. e and f. Rod-shaped particle. g. C-shaped particle. h. Cross-section through g, cutting the nucleocapsid twice. (From Weiss, M. and Horzinek, M.C., *Arch. Virol.*, 92, 1, 1987. With permission.)

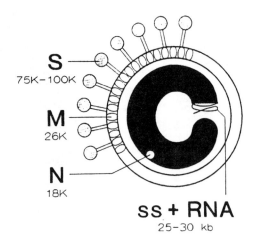

FIGURE 43

A more recent diagram indicating the location of M, N, and S (peplomer) proteins. (From Snijder, E.J. and Horzinek, M.C., *J. Gen. Virol.*, 74, 2305, 1993. With permission.)

CORONAVIRIDAE

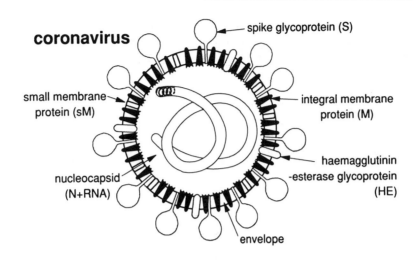

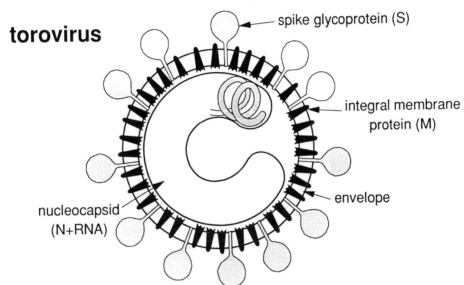

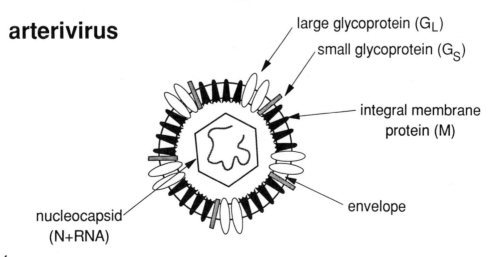

FIGURE 44

Comparative structure of corona-, toro-, and arteriviruses. (From Cavanagh, D., Brien, D.A., Brinton, M., Enjuanes, L., Holmes, K.V., Horzinek, M.C., Lai, M.M.C., Laude, H., Plagemann, P.G.W., Siddell, S., Spaan, W.J.M., Taguchi, R., and Talbot, P.J., *Arch. Virol.*, 135, 227, 1994. With permission.)

dsDNA, circular
Cubic, naked
Bacteria

VIII. CORTICOVIRIDAE

This family includes a single virus of marine origin, *Alteromonas* bacteriophage PM2. Particles are naked icosahedra with brush-like spikes on vertices. The capsid is multilayered and consists of 2 protein shells and a lipid bilayer (13% of particle weight) located in-between. Protein IV may be a transcriptase. The DNA of phage PM2 is frequently used as a molecular standard.

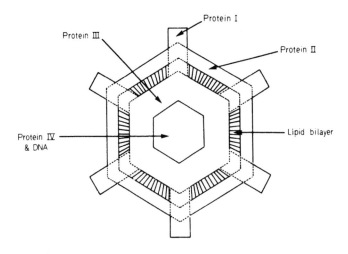

FIGURE 45
Section of bacteriophage PM2. The lipid bilayer consists of phosphatidylglycerol outside and phosphatidylethanolamine inside. (From Schäfer, R., Hinnen, R., and Franklin, F.M., *Eur. J. Biochem.*, 50, 15, 1974. With permission.)

dsRNA, 3 segments
Cubic, enveloped
Bacteria

IX. CYSTOVIRIDAE

The only member is *Pseudomonas* bacteriophage ø6, an enveloped virus of about 75 nm in diameter. The flexible envelope contains an icosahedral capsid with three molecules of dsRNA and a dodecahedral RNA polymerase complex. Phage ø6 is antigenically complex and the only bacterial virus with dsRNA.

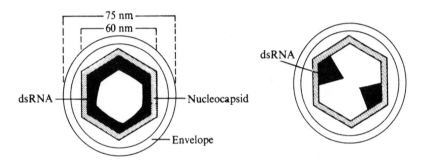

FIGURE 46

Bacteriophage ø6 sectioned in the plane of (a) and perpendicular to (b) the nucleic acid ring. The phage RNA is in close contact with the inner side of the capsid and leaves a translucent area at the center of the particle (the location of RNA polymerase?). (From Gonzalez, C.F., Langenberg, W.G., Van Etten, J.L., and Vidaver, A.K., *J. Gen. Virol.*, 35, 353, 1977. With permission.)

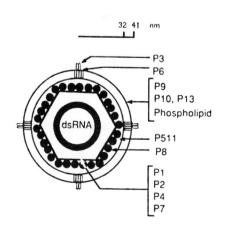

FIGURE 47

Three RNA molecules form a toroidal coil inside a dodecahedral structure composed of four proteins. P8 is the main capsid protein. The envelope is composed of phospholipid and three proteins. P3 and P6 are needed for fixation. (From Mindich, L., *Adv. Virus Res.*, 35, 137, 1988. With permission.)

ssRNA, + sense
Cubic, enveloped
Vertebrates, invertebrates

X. FLAVIRIDAE

This is a large family with three genera, *Flavivirus* (group B arboviruses including the agents of yellow fever and dengue), *Pestivirus* (pathogens of domestic animals), and human hepatitis C virus. Particles are enveloped, spherical, and 40 to 60 nm in diameter. Envelopes carry transmembrane protein spikes; capsids are icosahedra.

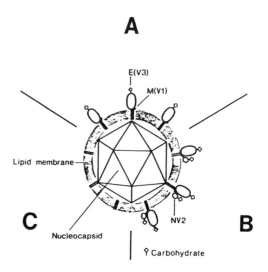

FIGURE 48
Structural components of a flavivirus. A. Mature particle. B. Immature intracellular particle. C. Pronase-treated virus. (From Heinz, F.X., *Adv. Virus Res.,* 31, 103, 1986. With permission.)

ssDNA, circular
Cubic, naked
Plants

XI. GEMINIVIRIDAE

One of the few plant virus groups with DNA. Particles consist of 2 incomplete icosahedra with 22 capsomers (T=1). There are three genera: members of one genus package two types of ssDNA molecules and have thus bipartite genomes.

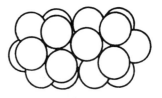

FIGURE 49
Two icosahedra are joined by loss of a morphological subunit. (From Stevens, W.A., *Virology of Flowering Plants,* Blackie & Sons, Glasgow, 1983, 89. With permission.)

ds(ss)DNA, circular
Cubic, enveloped
Vertebrates

XII. HEPADNAVIRIDAE

This small family consists of two genera of narrow host-specific viruses infecting mammals and birds. Particles are about 45 nm in diameter and consist of an envelope, an icosahedral capsid (T=3), and a single molecule of partially double-stranded DNA. Human hepatitis B viruses are named "Dane particles" after their discoverer. The blood of infected people frequently contains large quantities of round or filamentous particles (22 nm in diameter, up to 500 nm in length) that are chemically identical to the viral envelope or the "hepatitis B surface antigen" (HBsAg).

Human hepatitis B virus and associated structures

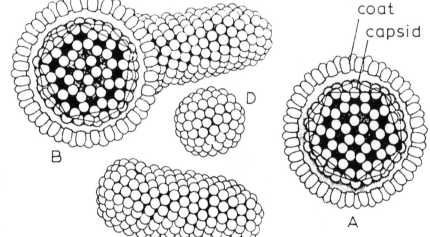

FIGURE 50

A. Normal isometric virus with icosahedral capsid or "core" surrounded by a coat of HBsAg. B. Tadpole-shaped virus with elongated coat. C and D. Filamentous and spherical assemblies of excess coat protein. (From Stannard, L.M., *Animal Virus Structure,* Nermut, M.V. and Steven, A.C., Eds., Elsevier, Amsterdam, 1987, 361. With permission.)

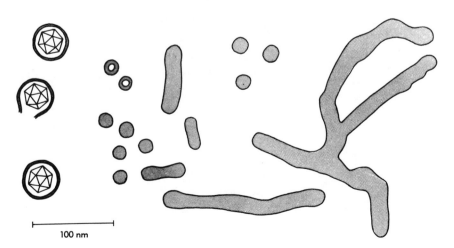

100 nm

FIGURE 51

A more naturalistic view. Some HBsAG particles are hollow. The structure at right is probably the most bizarre HBsAG particle observed. (By H.-W. Ackermann and M. Côté after Reference 58.)

HEPADNAVIRIDAE

Hepatitis B virus and its components

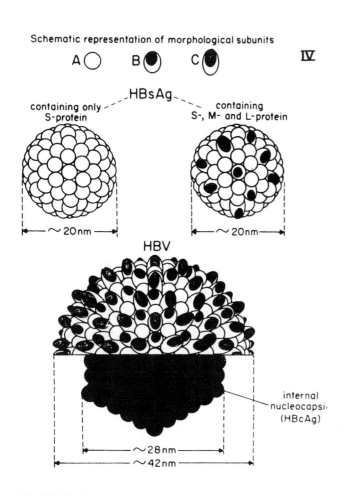

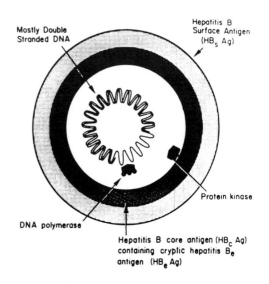

FIGURE 53
(From Lutwick, L.I., *Textbook of Human Virology*, 2nd ed., Belshe, R.B., Ed., Mosby-Year Book, St. Louis, 1991, 719. With permission.)

FIGURE 52
Composition of subunits and assembled HBs and virus particles. Individual subunits contain S protein only (A), M protein (B), or L protein (C). The lower half of the virus particle shows the nucleocapsid. Original in color. (From Neurath, A.R., Jameson, B.A., and Huima, T., *Microbiol. Sci.*, 4, 48, 1987. With permission.)

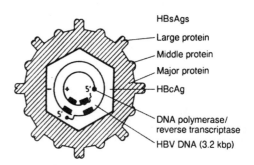

FIGURE 54
(From Blum, H.E., *Intervirology*, 35, 40, 1993. With permission of S. Karger AG, Basel.)

XIII. HERPESVIRIDAE

This large and diversified family is found in all types of vertebrates and in mollusks, and comprises many human pathogens. Members tend to cause latent infections. Viruses are 120 to 200 nm in diameter and morphologically complex. They consist of a loose envelope provided with spikes, a capsid with 150 hexameric and 12 pentameric hollow capsomers (T=16), and a fibrous "tegument" surrounding the capsid. The capsid contains a toroidal body (the DNA?) and a protein cylinder. There are three subfamilies according to DNA content and structure:

1. *Alphaherpesvirinae:* Herpes simplex and varicella viruses, others.
2. *Betaherpesvirinae:* Cytomegalovirus, others.
3. *Gammaherpesvirinae:* Epstein-Barr virus, others.

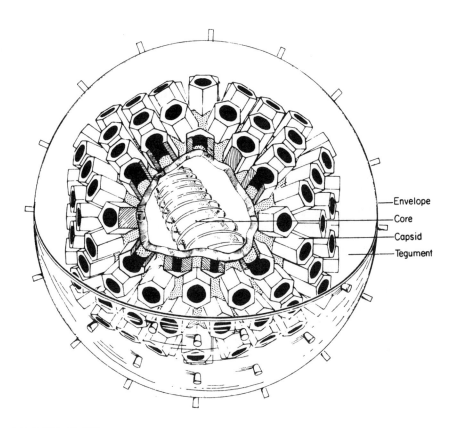

Envelope
Core
Capsid
Tegument

FIGURE 55
Particle showing envelope with surface projections, tegument, icosahedral capsid with 162 hollow hexagonal or pentagonal capsomers, and DNA core. (From Longson, M., *Principles and Practice of Clinical Virology,* 2nd ed., Zuckerman, A.J., Banatvala, J.E., and Pattison, J.E., Eds., 1990, 3. Reprinted by permission of John Wiley & Sons, Ltd., Chichester, England.)

HERPESVIRIDAE

Cutaway diagrams of a herpesvirus and its nucleocapsid

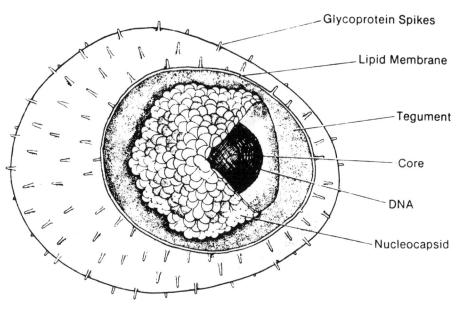

FIGURE 56

(From Hay, J., Roberts, C.R., Ruyechan, W.T., and Steven, A.C., *Animal Virus Structure*, Nermut, M.V. and Steven, A.C., Eds., Elsevier, Amsterdam, 1987, 391. With permission.)

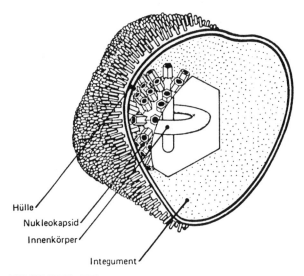

FIGURE 57

(With the kind permission of Blackwell Wissenschafts-Verlag GmbH. Taken from Horzinek, C., *Kompendium der allgemeinen Virologie*, 2nd ed., Paul Parey, Berlin, 1985, 49.)

HERPESVIRIDAE

Cross-sections of herpesviruses

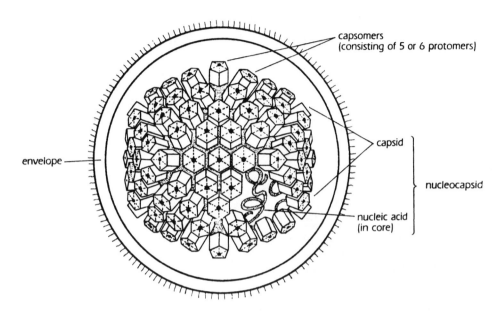

FIGURE 58

Particle with icosahedral capsid. (From Doane, F.W. and Anderson, N., *Electron Microscopy in Diagnostic Virology, A Practical Guide and Atlas,* 1987, 48. Reprinted with the permission of Cambridge University Press, New York.

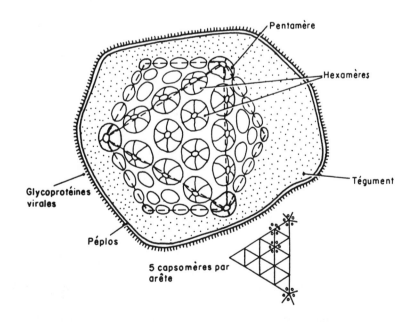

FIGURE 59

Particle with octahedral capsid! A regrettable error in an otherwise fine diagram (author's note). (From Huraux, J.M., Nicolas, J.C., and Agut, H., *Virologie,* 1985, 69, Flammarion, Paris. With permission.)

HERPESVIRIDAE

Herpesvirus morphology

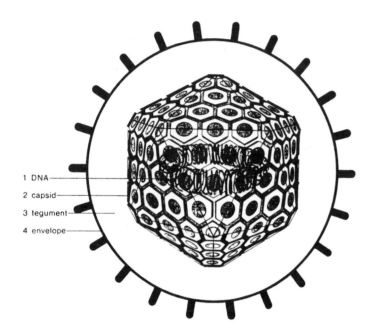

1 DNA
2 capsid
3 tegument
4 envelope

FIGURE 60

Complete virus particle; original in color. (From McKendrick, G.D.W. and Sutherland, S., *An Introduction to Herpes Virus Infections,* Wellcome Foundation, London, 1983, 10. With permission.)

FIGURE 61

Herpesvirus capsid viewed along a two-fold symmetry axis. It has 12 pentagonal capsomers located at the apexes and 150 hexagonal capsomers on faces and sides. (From Horne, R.W., *The Structure and Function of Viruses,* 1978, 7. © 1978 by R.W. Horne. Reproduced by permission of Edward Arnold, Ltd., London.)

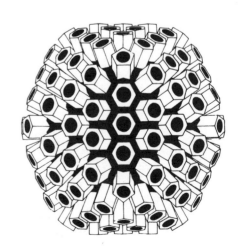

HERPESVIRIDAE

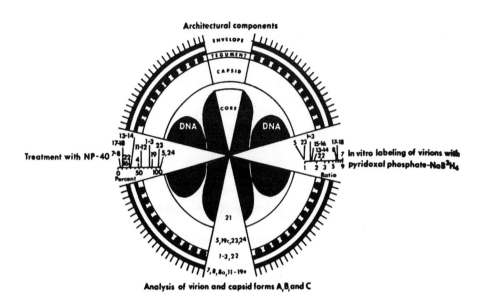

FIGURE 62

The HSV-1 virus and its architectural components. Bottom segments show polypeptides present in the intact virion, empty (A) and full (B) capsids, and viral membranes. Left segment shows the possible order of polypeptides in envelope and tegument based on stripping experiments with the detergent NP-40. Numerals are percentages of polypeptides remaining after treatment. Right segment shows another possible order of polypeptides derived from accessibility to labeling by tritiated borohydride reduction of Schiff's bases formed between polypeptides and pyridoxal phosphate. (From Roizman, B. and Furlong, D., *Comprehensive Virology*, Vol. 3, Fraenkel-Conrat, H. and Wagner, R.R., Eds., Plenum Press, New York, 1974, 229. With permission.)

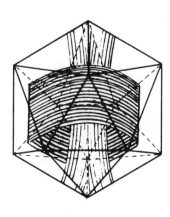

FIGURE 63

Nucleocapsid of Marek's disease virus showing location of the DNA-containing torus. (From Nazerian, K., *J. Virol.*, 13, 1148, 1974. With permission.)

dsDNA, circular
Helical, naked
Bacteria

XIV. INOVIRIDAE

The family includes 2 genera with about 50 phages. Members of the genus *Inovirus* are long, rigid, or flexuous filaments of 760 to 2000 × 6 to 8 nm and occur in a group of related Gram-negative bacteria (enterobacteria, pseudomonads, xanthomonads, vibrios). They belong to two structural classes. Phages of the genus *Plectrovirus* are short rods and occur in mycoplasmas only.

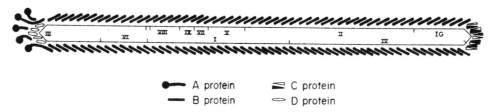

● A protein   C protein
— B protein   D protein

FIGURE 64

Phages of the *Inovirus* genus, exemplified by coliphage fd, consist essentially of DNA, the terminal A protein, and the major coat protein B which forms a double-layered sheath around the DNA. B protein subunits are aligned more or less parallel to the length of the particle and overlap each other. The diagram shows the relative position of protein molecules and does not suggest any interactions between proteins or between proteins and DNA. (From Webster, R.E., Grant, R.A., and Hamilton, L.A.W., *J. Mol. Biol.,* 152, 357, 1981. With permission.)

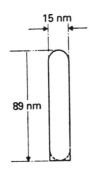

FIGURE 65

Mycoplasma phage L1 (*Plectrovirus* genus). (From Horne, R.W., *The Structure and Function of Viruses,* 1978, 49. © R.W. Horne. Reproduced by permission of Edward Arnold, Ltd., London.)

FIGURE 66

L1 phage diagram showing two-start-helix of hexagonally arranged subunits. (By G. Larose after a model in Reference 70.)

INOVIRIDAE

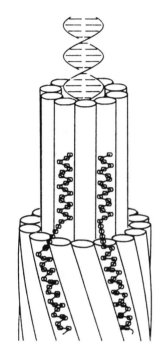

FIGURE 67

Class I inovirus structure. The protein coat is depicted as consisting of pentamers of α-helical subunits. Neighboring subunits interlock so that there are ten subunits in the inner layers and five in the outer layer. Separate domains of the same subunit contribute to both layers. The right figure shows how the DNA fits into the protein coat. (From Webster, R.E. and Lopez, J., *Virus Structure and Assembly*, 235, Casjens, S., Ed. © 1985 Jones and Bartlett Publishers, Boston. Reprinted by permission.)

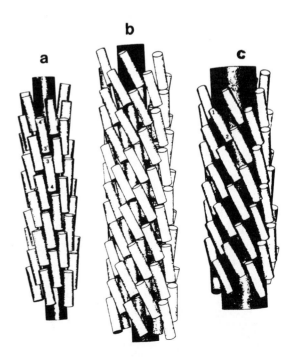

FIGURE 68

Class II inovirus structure, Pf1 phage. The diagram shows (a) repeats of the inner layer of coat protein, (b) two repeats of the outer layer, and (c) 2.5-Å repeats of both layers. The a-helical segments are shown as cylinders of 3×0.9 nm. The DNA is represented as a dark cylinder of 4 nm in diameter. (From Makowski, L., Caspar, D.L.D., and Marvin, D.A., *J. Mol. Biol.*, 140, 149, 1980. With permission.)

dsDNA, linear
Cubic, naked
Vertebrates, invertebrates

XV. IRIDOVIRIDAE

This is a highly diversified virus group with five genera that is found in cold-blooded animals (fish, frogs, insects, worms). Typical particles are naked icosahedra of 125 to 300 nm in diameter and complex, multilayered capsids with 5 to 9% lipids. The number of capsomers has been estimated at 812, 1472, or 1562. Some members possess an envelope derived from the plasma membrane of the host cell.

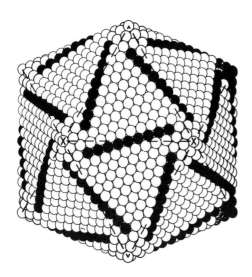

FIGURE 69

Outer shell of *Sericesthis iridescent* virus, thought to consist of skewed triangles with 1562 capsomers (handedness is arbitrary). (From Wrigley, N.G., *J. Gen. Virol.*, 5, 123, 1969. With permission.)

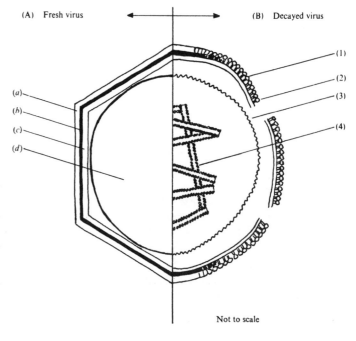

FIGURE 70

Lymphocystis virus of fish. A. Fresh virus: a, outer electron-transparent layer; b, middle electron-dense layer; c, inner electron-transparent layer; d, apparently structureless core. B. Decayed virus: 1, stain-penetrated layers a and b showing knob-like subunits; 2, stalks; 3, boundary of core; 4, internal tubules of 13 × 4 nm. (From Madeley, C.R., Smail, D.A., and Egglestone, S.I., *J. Gen. Virol.*, 40, 421, 1978. With permission.)

IRIDOVIRIDAE

Two views of Lymphocystis virus of fish by the same authors

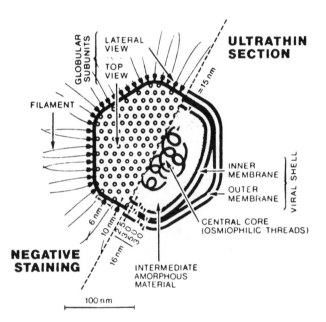

FIGURE 71
Virus according to negative staining and thin sectioning. (From Berthiaume, L., Alain, R., and Robin, J., *Virology,* 135, 10, 1984. With permission.)

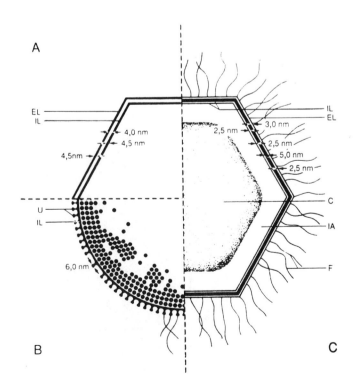

FIGURE 72
Viruses in glutaraldehyde-fixed (A) and unfixed (B) negatively stained preparations and in thin section (C). Note in B the larger size and more spherical outline of particle. C, core; EL, external layer; F, fibrils; IA, intermediate area; IL, internal layer, U, surface units. (From Heppell, J. and Berthiaume, L., *Arch. Virol.,* 125, 215, 1992. With permission.)

IRIDOVIRIDAE

Two views of frog virus 3 (FV3)

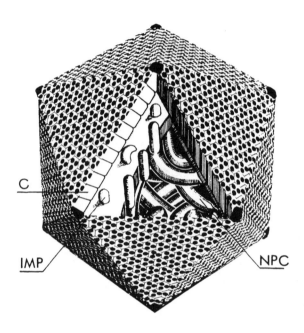

FIGURE 73

Capsomers (C) are shown as trimers on a hexagonal net (T=133). IMP, inner membrane with intramembranous particles. NPC, nucleoprotein complex represented as a convoluted filamentous structure. (From Darcy, F. and Devauchelle, G., *Animal Virus Structure*, Nermut, M.V. and Steven, A.C., Eds., Elsevier, Amsterdam, 1987, 407. With permission.)

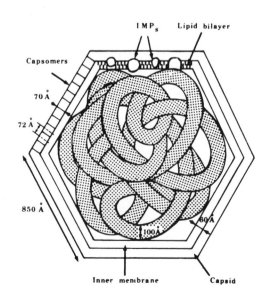

FIGURE 74

Section with dimensions of main components. IMPs, intramembrane particles. (From Darcy-Tripier, F., Nermut, M.V., Braunwald, J., and Williams, L.D., *Virology*, 138, 287, 1984. With permission.)

ssRNA, + sense
Cubic, naked
Bacteria

XVI. LEVIVIRIDAE

This group of about 90 ssRNA phages is restricted to enterobacteria, pseudomonads, and caulobacters, two genera. Particles are naked icosahedra of about 24 nm in diameter and have 32 capsomers (T=3). Leviviruses have structural but apparently no phylogenetic relationships to picornaviruses and ssRNA plant viruses with cubic symmetry.

Two views of complete particles

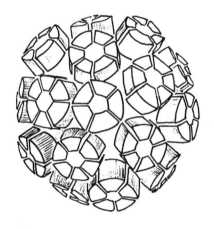

FIGURE 75
(By L. Berthiaume after a model in Reference 79.)

FIGURE 76
(From Paranchych, W., *RNA Phages*, Zinder, N.D., Ed., Cold Spring Harbor Laboratory, Cold Spring Harbor, NY, 1975, 85. With permission.)

FIGURE 77
Phage MS2. a. Icosahedron with location of symmetry elements on triangular facets. b. Position of subunits on a facet and neighboring subunits. (Reprinted with permission from *Nature*, 344, 36. © 1990 Macmillan Magazines Ltd., London)

dsDNA, linear
Helical, enveloped
Bacteria

XVII. LIPOTHRIXVIRIDAE

This very small phage group restricted to archaebacteria of the *Thermoproteales* branch. Only one member, TTV1, has been studied in detail. Particles are rods of about 400 × 40 nm and consist of an envelope, a helical nucleoprotein core, and protrusions at each end. Similar particles of different length are known, but have not been characterized.

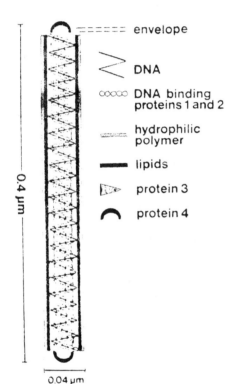

FIGURE 78

Thermoproteus phage TTV1 according to fragmentation experiments. A DNA-protein complex is associated with protein P3 to form a tube, which is covered by P4 caps and surrounded by a lipid-containing unit membrane. (From Zillig, W.R., Reiter, H.-D., Palm, P., Gropp, P., Neumann, H., and Rettenberger, M., *The Bacteriophages*, Vol. 1, Calendar, R., Ed., Plenum Press, New York, 1987, 517. With permission.)

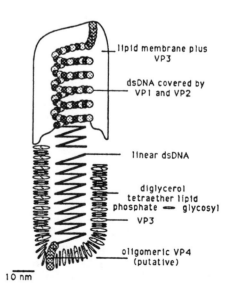

FIGURE 79

Phage TTV1. The upper half of the diagram shows DNA associated with DNA-binding proteins; the lower half shows superhelical DNA without covering protein and the composition of the coat. The central part of the virus is not shown. (From Zillig, W., *Virus Taxonomy. The Classification and Nomenclature of Viruses. Sixth Report of the International Committee on Taxonomy of Viruses, Arch. Virol.*, 1995. With permission.)

ssDNA, circular
Cubic, naked
Bacteria

XVIII. MICROVIRIDAE

This group of about 40 phages is divided into four genera, occurring in enterobacteria, bdellovibrios, chlamydias, and spiroplasmas. Particles are naked icosahedra of 25 to 27 nm in diameter and consist of 12 capsomers (T=1); some members have apical spikes.

Two representations of phage øX174

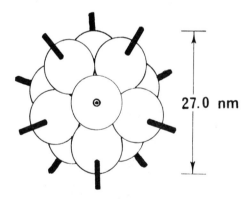

27.0 nm

FIGURE 80
An early model (1974). Capsomers are drawn as spheres with a central spike projecting from the center. (From Horne, R.W., *Virus Structure*, Academic Press, New York, 1974, 7. With permission.)

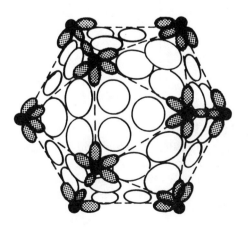

FIGURE 81
A slightly later diagram (1978). White circles represent the major capsid protein, VPF. The apical spikes are composed of VPG major capsid protein, VPF. The apical spikes are composed of VPG (cross-hatched) and a single molecule of VPH (black). (From Hayashi, M., *The Single-Stranded DNA Phages*, Denhardt, D.T., Dressler, D., and Ray, D.S., Eds., Cold Spring Harbor Laboratory, Cold Spring Harbor, NY, 1978, 531. With permission.)

ssRNA, – sense, 1 segment
Helical, enveloped
Vertebrates, invertebrates, plants

XIX. MONONEGAVIRALES

So far this is the only order established in virology; defined by the presence of a single molecule of negative-sense ssRNA, an envelope, a helical nucleocapsid, and virion-associated RNA polymerase. The order includes three families.

Vertebrates

A. PARAMYXOVIRIDAE

Viruses are mammal-specific and divided into two subfamilies *(Paramyxovirinae, Pneumovirinae)* and four genera. They include important human and animal pathogens, e.g., measles, mumps, rinderpest, canine distemper, Newcastle disease, and respiratory syncytial virus. Particles are roughly spherical and 150 to 300 nm in size, consisting of an envelope with spikes (transmembrane glycoproteins) and a single, more or less rigid nucleocapsid of 13 to 18 nm in diameter. Members of the genus *Paramyxovirus* possess spikes with neuraminidase activity. Until 1970, paramyxoviruses and influenza viruses *(Orthomyxoviridae)* were grouped together as "myxoviruses".

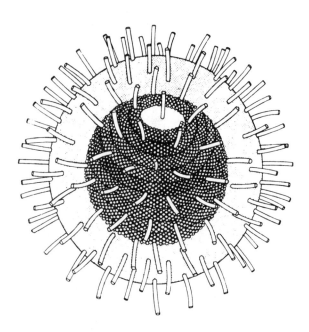

FIGURE 82
"Myxovirus" showing envelope with regularly spaced surface projections (1974). The nucleoprotein is shown as a coiled cylinder. (From Horne, R.W., *Virus Structure*, Academic Press, New York, 1974, 28. With permission.)

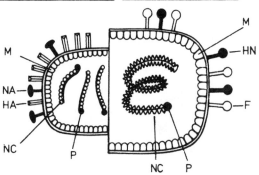

FIGURE 83
Comparative structure of ortho- and paramyxoviruses. F, fusion protein; HA, hemagglutinin; HN, neuraminidase; NC, nucleocapsid; P, membrane protein. (From Rott, R., *Arch. Virol.*, 59, 285, 1979. With permission.)

PARAMYXOVIRIDAE

Paramyxovirus architecture

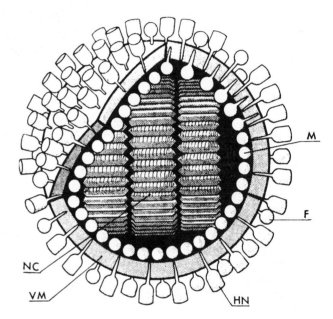

FIGURE 84
F, fusion spike; HN, hemagglutinin-neuraminidase spike; M, M protein shown with and without transmembrane side chain; NC, nucleocapsid; VM, viral membrane. (From Scheid, H., *Animal Virus Structure*, Nermut, M.V. and Steven, A.C., Eds., Elsevier, Amsterdam, 1987, 233. With permission.)

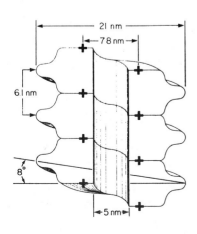

FIGURE 85
Segment of measles virus nucleocapsid indicating the calculated position of the RNA (crosses). (From Lund, G.A., Tyrrell, D.L.J., Bradley, R.D., and Scraba, D.G., *J. Gen. Virol.*, 65, 1535, 1984. With permission.)

PARAMYXOVIRIDAE

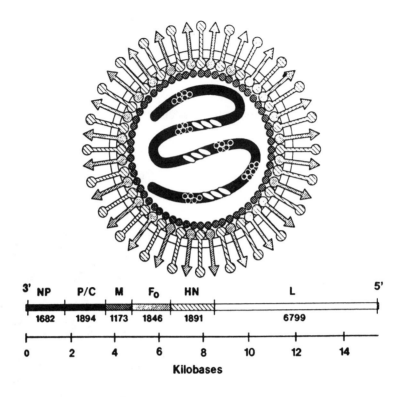

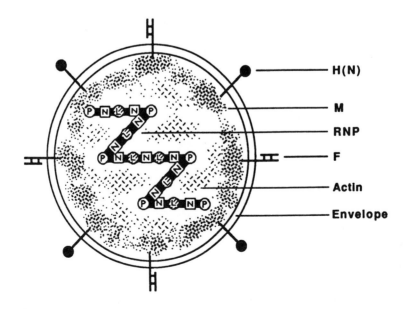

FIGURE 86

Structure and genetic map of Sendai virus. Symbols of viral proteins do not indicate their actual shape. The nucleocapsid filament contains RNA, structure units NP (black), and auxiliary proteins L (white ovals) and P (circles). The envelope is penetrated by fusion protein F (pointed tips) and attachment glycoprotein HN (round tips). M, inner envelope protein. (From Kingsbury, D.W., *Virology*, 2nd ed., Vol. 1, Fields, B.N. and Knipe, D.M., Eds., Raven Press, New York, 1990, 945. With permission.)

FIGURE 87

Structure of a paramyxovirus. F, fusion glycoprotein; H(N) hemagglutinin (neuraminidase) glycoprotein; L, large protein; M, matrix (membrane); N, nucleoprotein; P, phosphoprotein; RNP, ribonucleoprotein. (From Vainionpää, R., Marusyk, R., and Salmi, A., *Adv. Virus Res.*, 37, 211, 1989. With permission.)

B. RHABDOVIRIDAE

This large and diversified family has an exceptional host range and occurs in vertebrates, insects, and plants. It is divided into five genera, notably *Lyssavirus* (rabies) and *Vesiculovirus* (vesicular stomatitis). Various members replicate in both vertebrates and insects or in insects and plants. Particles are bullet-shaped (vertebrate and insect viruses) or bacilliform (many plant viruses) and measure 100 to 430 × 50 to 100 nm. They consist of an outer envelope with small spikes and a tubular nucleocapsid about 50 nm in diameter that is formed by a coiled filament and gives the particle a striated appearance.

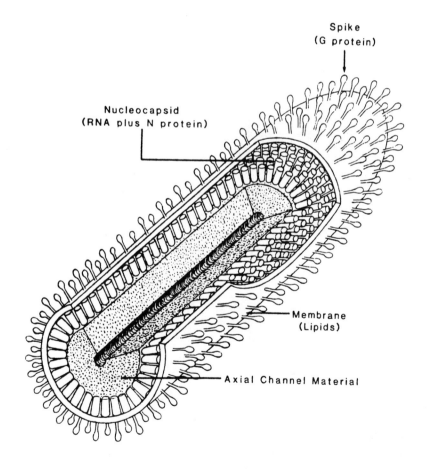

FIGURE 88

Vesicular stomatitis virus (VSV) showing nucleocapsid (N, L, and S proteins), envelope, spikes (G), and axial channel (filled with M protein?). (From Brown, J.C. and Newcomb, W.W., *Animal Virus Structure*, Nermut, M.V. and Steven, A.C., Eds., Elsevier, Amsterdam, 1987, 199. With permission.)

RHABDOVIRIDAE

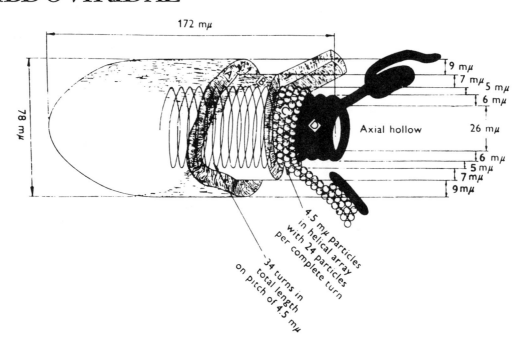

FIGURE 89

Vesicular stomatitis virus (1964). A. Surface elements. B. Envelope. C. Helical "capsomers". D. Core material. (From Bradish, C.J. and Kirkham, J.B., *J. Gen. Microbiol.*, 44, 359, 1966. With permission.)

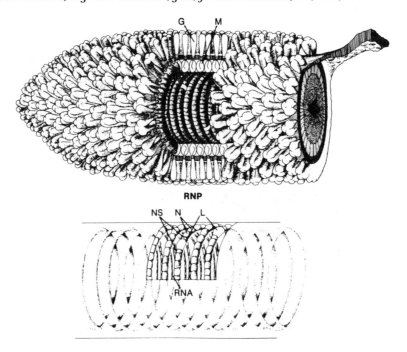

FIGURE 90

Rabies virus (1988) with surface glycoprotein projections (G) shown as trimers, envelope, and nucleocapsid surrounded by matrix (M) protein. L, transcriptase; N, nucleoprotein; NS, phosphoprotein; RNP, ribonucleoprotein complex. The protruding "tail" symbolizes the frequently irregular shape of viruses. (From Wunner, W.H., Larson, J.K., Dietzschold, B., and Smith, C.L., *Rev. Infect. Dis.*, 10(Suppl. 4), 771. © 1988 by University of Chicago Press, Chicago, IL, with permission.)

RHABDOVIRIDAE

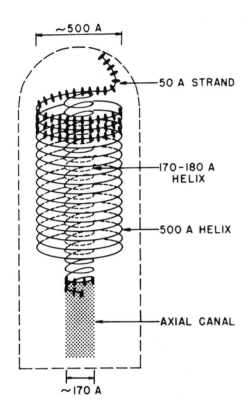

FIGURE 91

Vesicular stomatitis virus nucleocapsid. The model assumes the existence of an inner and an outer helix derived from the same strand and extending in about 35 turns along the entire particle. (From Simpson, R.W. and Hauser, R.E., *Virology*, 29, 654, 1966. With permission.)

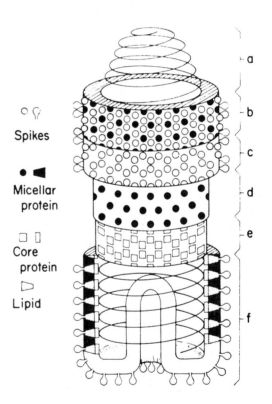

FIGURE 92

Rabies virus. a. Decreasing turns of nucleocapsid. b. Surface view showing the relative location of spikes and micellar proteins. c. Spikes. d. Micellar protein. e. Internal membrane-like layer (core protein). f. Section of particle showing relation of lipids to the micellar layer. The spikes may extend farther into the envelope. A portion of the latter extends into the central cavity of the nucleoprotein helix. For clarity, the diagram does not show the subunits of the nucleocapsid and the subunit structure of the envelope at the planar end of the particle. The nucleocapsid is arbitrarily represented as a right-handed helix. (From Vernon, S.K., Neurath, A.R., and Rubin, B.A., *J. Ultrastruct. Res.*, 41, 29, 1972. With permission.)

RHABDOVIRIDAE

Plant rhabdoviruses

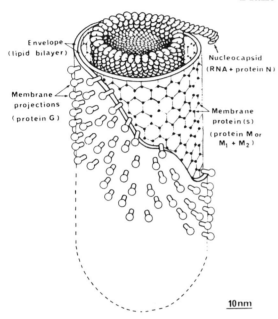

FIGURE 93
Complete, partly sectioned particle. (Reprinted from Francki, R.I.B. and Randles, J.W., *Rhabdoviruses,* Vol. 3, Bishop, D.H.L., Ed., 1980, 135, CRC Press, Boca Raton, FL.)

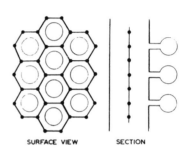

FIGURE 94
Model of envelope showing hexameric arrangement of subunits and section of surface projections. (From Francki, R.I.B., *Adv. Virus Res.,* 18, 257, 1973. With permission.)

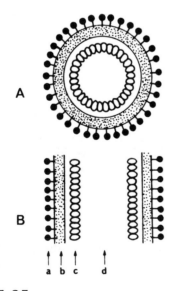

FIGURE 95
Transverse (A) and longitudinal sections (B) showing surface projections (a), envelope (b), nucleocapsid (c), and central channel (d). (From Francki, R.I.B., *Adv. Virus Res.,* 18, 257, 1973. With permission.)

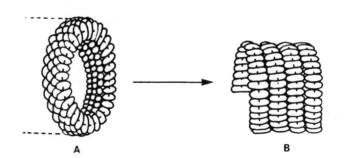

FIGURE 96
Configuration of nucleocapsid in the intact particle (A) and after removal of the envelope (B). (From Francki, R.I.B., *Adv. Virus Res.,* 18, 257, 1973. With permission.)

RHABDOVIRIDAE

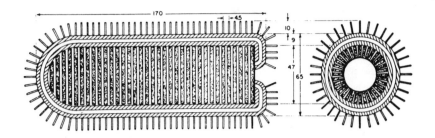

FIGURE 97

Length and cross-sections of VSV; dimensions in nanometers. The nucleo-protein appears at left as a series of transverse striations with a spacing of 4.5 nm. (From Howatson, A.F., *Adv. Virus Res.,* 16, 195, 1970. With permission.)

Vertebrates

C. FILOVIRIDAE

Another small family with only two members, Marburg and Ebola viruses, includes agents of severe hemorrhagic fevers in humans. Particles are extremely pleomorphic, often appearing U-shaped, branched, or circular. Preferential dimensions of simple elongated particles are 800 to 1000 × 80 nm. As rhabdoviruses, they have an envelope and a tubular, striated nucleocapsid.

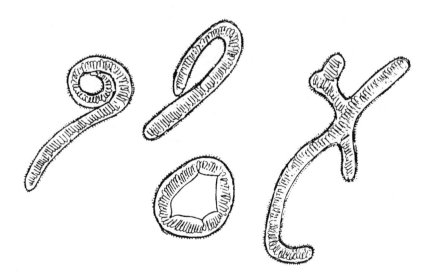

FIGURE 98

Polymorphic virus particles. (From Mammette, A., *Virologie médicale à l'usage des étudiants en médecine,* 14th ed., Editions C. et R., La Madeleine, France, 1992, 350. With permission.)

ssRNA, 8 segments
Helical, enveloped
Vertebrates

XX. ORTHOMYXOVIRIDAE

This small family with three genera, found in mammals, birds, and ticks includes the influenza viruses. Particles are enveloped, usually rounded, and 80 to 120 nm in diameter; filamentous forms occur. The envelope is studded with glycoprotein spikes. The genome is segmented and consists of six to eight nucleoprotein complexes accounting for frequent recombinations. Influenza A and B viruses contain eight RNA segments and have spikes with hemagglutinin and neuraminidase function. Influenza C viruses contain seven molecules of RNA and have a single type of spikes.

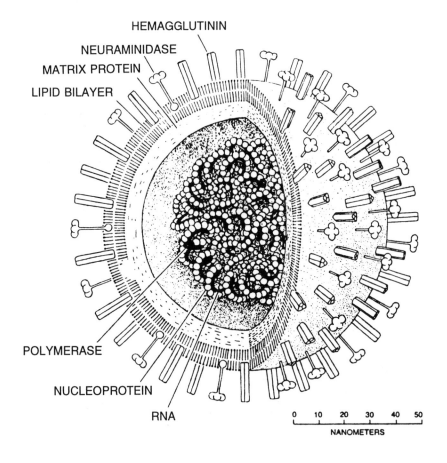

FIGURE 99

Cutaway diagram of influenza virus. Hemagglutinin and neuraminidase spikes are embedded in a lipid bilayer. The core consists of a nucleoprotein complex (RNA, nucleoprotein, RNA polymerase). The RNA is divided into eight segments of different length. (From Kaplan, M.M. and Webster, R.G., *Sci. Am.*, 273(6), 88. © 1977 Scientific American, Inc., New York. All rights reserved.)

ORTHOMYXOVIRIDAE

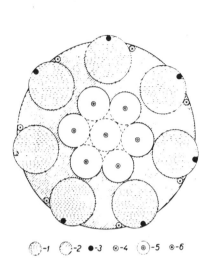

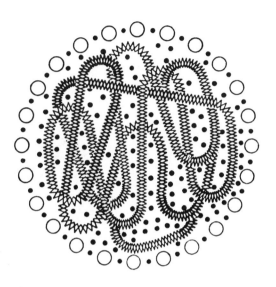

FIGURE 100

The first diagram of an influenza virus (1959). 1. Hemagglutinin protein. 2. Lipids. 3. Virus enzyme. 4. Carbohydrates. 5. S- or g-antigen, a nucleoprotein. 6. Nucleic acid. (From Blaskovic, D., *Acta Virol.*, 3(Suppl.), 7, 1959. With permission.)

FIGURE 101

Early diagram (1962) suggesting how lipids (black dots) may hold the helical nucleoprotein and the hemagglutinin (white circles) together. (From Kates, M., Allison, A.C., Tyrell, D.A.J., and James, A.T., *Cold Spring Harbor Symp. Quant. Biol.*, 27, 193, 1962. With permission.)

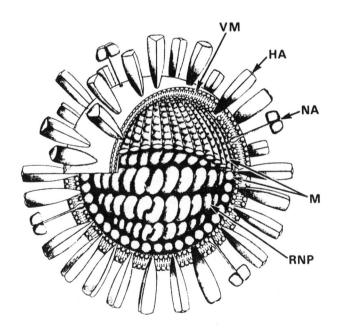

FIGURE 102

Recent representation (1987) showing the viral membrane (VM), hemagglutinin (HA) and neuraminidase (NA) spikes, M protein, and the ribonucleoprotein core (RNP). Structural components are drawn to scale, but the viral particle corresponds to 50 nm only (about half of the normal size of the virus). (From Oxford, J.S. and Hockley, D.J., *Animal Virus Structure*, Nermut, M.V. and Steven, A.C., Eds., Elsevier, Amsterdam, 1987, 213. With permission.)

ORTHOMYXOVIRIDAE

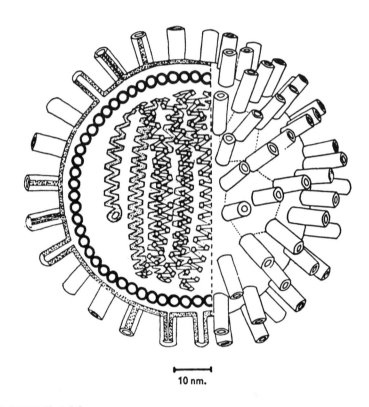

10 nm.

FIGURE 103

Morphology of influenza A and B viruses. (From Apostolov, K. and
Flewett, T.H., *J. Gen. Virol.*, 4, 365, 1969. With permission.)

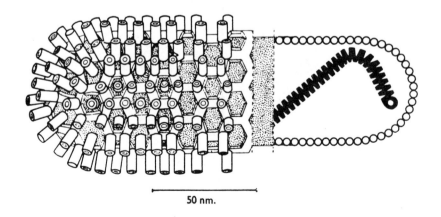

50 nm.

FIGURE 104

Morphology of the influenza C filament. (From Apostolov, K. and Flewett,
T.H., *J. Gen. Virol.*, 4, 365, 1969. With permission.)

ORTHOMYXOVIRIDAE

Sections of influenza viruses

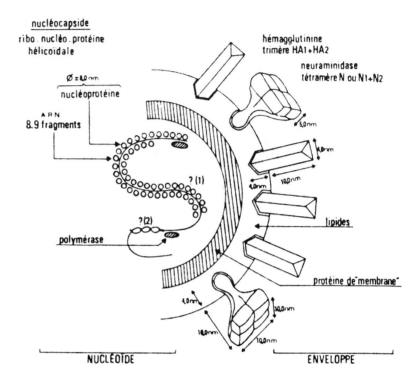

FIGURE 105
(From Aymard, M., *Virologie médicale,* Maurin, J., Ed., Flammarion, Paris, 1985, 448. With permission.)

(1) génome segmenté
(2) génome sous forme "bicaténaire"

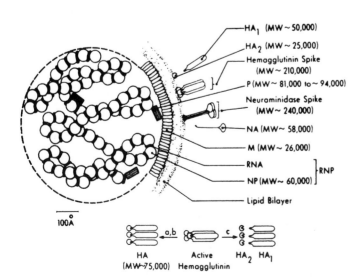

FIGURE 106
Model and molecular weights of components. HA, hemagglutinin; M, matrix protein; NA, neuraminidase subunit; NP, nucleoprotein; P, RNA polymerase. (From Schulze, I.T., *Adv. Virus Res.,* 18, 1, 1973. With permission.)

ORTHOMYXOVIRIDAE

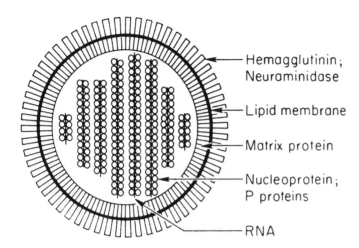

FIGURE 107
Influenza A virus with eight nucleoprotein complexes. (Reprinted from Palese, P. and Ritchey, M.D., *Virology and Rickettsiology,* Vol. 1, Hsiung, G.-D. and Green, R.H., Eds., 1978, 337, CRC Press, Boca Raton, FL.)

FIGURE 108
Individual nucleoprotein complex of influenza virus. Dark grooves suggest areas of a greater stain penetration. (From Compans, R.W., Content, J., and Duesberg, P.H., *J. Virol.,* 10, 795, 1972. With permission.)

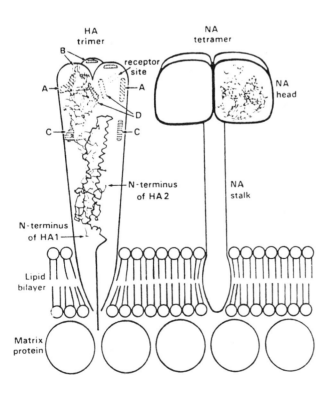

FIGURE 109
Hemagglutinin (HA) and neuraminidase (NA) spikes and their possible relationship with the lipid bilayer envelope and matrix protein. A to D, antigenic sites of hemagglutinin. HA spikes are trimers and NA spikes are tetramers. (From Oxford, J.S. and Hockley, D.J., *Animal Virus Structure,* Nermut, M.V. and Steven, A.C., Eds., 1987, 213, Elsevier, Amsterdam. With permission.)

dsDNA, circular
Cubic, naked
Vertebrates

XXI. PAPOVAVIRIDAE

Viruses tend to cause tumors or cell transformation *in vitro*. They infect mammals, birds, and reptiles and are classified into the genera *Papillomavirus* (papillomas, warts) and *Polyomavirus*. Particles are naked icosahedra of 40 to 55 nm in diameter. The capsid consists of 72 skewed, hollow capsomers (T=7*l*).

Four interpretations of papovavirus structure

FIGURE 110
Polyomavirus model with 42 capsomers (1963). (Modified by H.-W. Ackermann from Reference 108.)

FIGURE 111
Papillomavirus with 72 skewed capsomers. (From Horne, R.W., *Virus Structure*, Academic Press, New York, 1974, 17. With permission.)

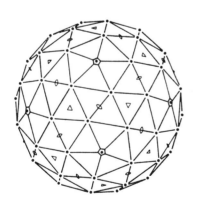

FIGURE 112
Polyomavirus representation with alternative view of a T=7*d* surface lattice (72 capsomers) and location of two-, three-, and fivefold axes of symmetry. (From Rayment, I., *Biological Macromolecules and Assemblies*, Vol. 1, 255, Jurnak, F.A. and McPherson, A., Eds., © 1984 by John Wiley & Sons, New York. Reprinted by permission.)

FIGURE 113
Sketch of polyomavirus. (Reprinted with permission from *Nature*, 178, 1453. © 1956 Macmillan Magazines Ltd., London.)

PAPOVAVIRIDAE

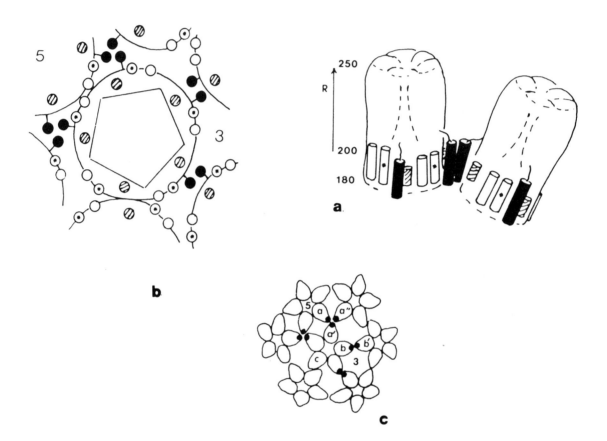

FIGURE 114

Structure of papillomavirus SV40 pentamers. a. Central cavity and rows of monomers. Each monomer contains four helices (white, starred, hatched, black). Numerals show radial positions of helices in virus particles. b. Six-coordinated pentamer seen from above. Alternative helix-helix packings are the basis for multiplicity of bonding interactions. Markings correspond to those in (a). Local interactions are twofold or threefold. Numerals 3 and 5 indicate icosahedral axes. c. Enlargement of the area around a six-coordinated pentamer. (From Harrison, S.C., *Virology*, 2nd ed., Vol. 1, Fields, B.N. and Knipe, B.M., Eds., Raven Press, New York, 1990, 37. With permission.)

ssDNA, linear
Cubic, naked
Vertebrates, invertebrates

XXII. PARVOVIRIDAE

Viruses are widely distributed in animals (mammals, birds, insects, crustacea) and form two subfamilies and six genera. Particles are naked icosahedra (T=1) and possibly the smallest of all viruses (18 to 25 nm in diameter). Members of the genus *Dependovirus* require the presence of adenoviruses or herpesviruses as helpers in replication.

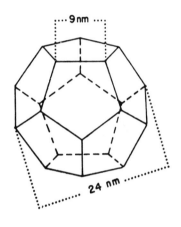

FIGURE 115

Densonucleosis virus represented as a dodecahedron; the only parvovirus diagram in the literature. (From Kurstak, E., Tijssen, P., and Garzon, S., *The Atlas of Insect and Plant Viruses,* Maramorosch, K., Ed., Academic Press, New York, 1977, 67. With permission.)

dsDNA, circular
Pleomorphic
Bacteria

XXIII. PLASMAVIRIDAE

This very small phage group with only two certain members is limited to mycoplasmas. Particles are spherical or pleomorphic, of about 80 nm (range 50 to 125 nm) in diameter, and consist of an envelope and a nucleoprotein granule without apparent structure.

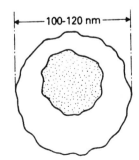

FIGURE 116

Mycoplasma virus L2. Particles are approximately spherical and have an envelope and an electron-dense central region. (From Horne, R.W., *The Structure and Function of Viruses,* 1978, 49. © R.W. Horne. Reproduced by permission of Edward Arnold, Ltd., London.)

ssRNA, + sense
Cubic, naked
Vertebrates, invertebrates

XXIV. PICORNAVIRIDAE

This large group includes five genera of mammal-specific viruses, notably such important human pathogens as polioviruses (genus *Enterovirus*) and the agent of hepatitis A (genus *Hepatovirus*). Particles are naked, without surface structures, and 28 to 30 nm in diameter. The capsid behaves as an assembly of 5 pentamers, combines features of the icosahedron and the dodecahedron, and consists of 60 subunits (T=1, pseudo T=3) of proteins VP1, VP2, and VP3. A fourth protein, VP4, cleaved from the precursor of VP2, is located inside the shell.

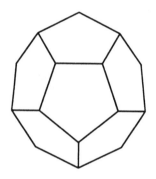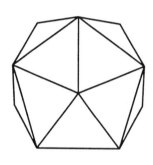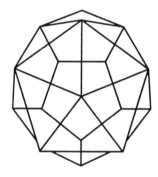

FIGURE 117
Poliovirus symmetry. A dodecahedron (a) superposed on an icosahedron (b) results in a multifaceted structure. (By H.-W. Ackermann and G. Larose, modified from Reference 109.)

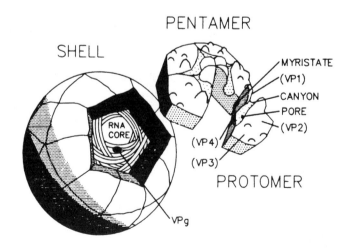

FIGURE 118
Exploded diagram of typical picornavirus showing myristate residues in center of pentamer. (From Rueckert, R.R., *Virology,* 2nd ed., Raven Press, 1990, 507. With permission.)

PICORNAVIRIDAE

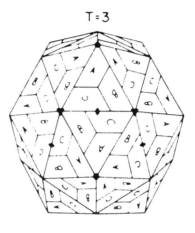

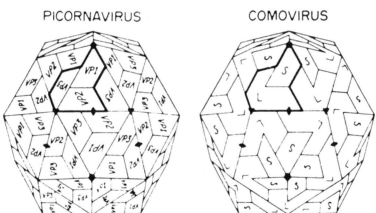

FIGURE 119

Comparative structure of a T=3 capsid, a picornavirus, and a comovirus (plants). Each trapezoid represents a ß-barrel. Each asymmetric icosahedral unit of the T=3 capsid contains three identical subunits (A, B, C). The asymmetric unit of the poliovirus (heavy outline) consists of three similar ß-barrels of different amino acid sequence (VP1, VP2, VP3). In the comovirus capsid, two ß-barrels are covalently linked to form a single-polypeptide, large-protein subunit (L); the small subunit (S) corresponds to VP1. The complete comovirus may be described as an assembly of 30 monomers (S) and 30 dimers (L). Picornaviruses and comoviruses have a similar gene order, and parts of their nonstructural proteins show significant sequence homology. The similarity of T=3, picornaviruses, and comoviruses raises the possibility of a common origin. (From Chen, Zh., et al., *New Aspects of Positive-Strand RNA Viruses,* Brinton, M.A. and Heinz, F.X., Eds., American Society for Microbiology, Washington, DC, 1990, 218. With permission.)

dsDNA, circular, segmented
Helical, enveloped
Invertebrates

XXV. POLYDNAVIRIDAE

This small group of viruses so far is found in endoparasitic wasps only and differs from all other DNA viruses by the presence of segmented genomes (several DNA molecules of variable size in each virion). Members of the *Ichnovirus* genus have a double envelope and one or two nucleocapsids of about 330 × 85 nm, resembling baculoviruses. The genus *Bracovirus* is characterized by a single envelope containing one or more nucleocapsids of 30 to 150 nm in length and often provided with tail-like appendages.

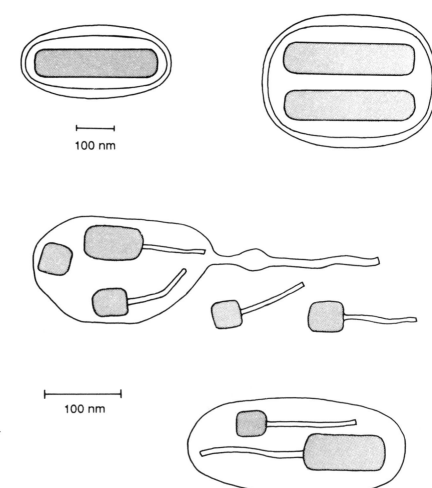

100 nm

100 nm

FIGURE 120

Ichnoviruses (above) and bracoviruses (below). Note the variable length of nucleocapsids and "tails" in the latter. The bar represents 100 nm. (By H.-W. Ackermann and M. Côté.)

dsDNA, linear
Helical, enveloped
Vertebrates, invertebrates

XXVI. POXVIRIDAE

This is one of the largest viral groups, occurring in a wide variety of mammals and birds (subfamily *Chordopoxvirinae*) and insects (*Entomopoxvirinae*) and includes 11 genera. Particles are brick shaped or ovoid, measure 220 to 450 × 140 to 260 × 110 to 200 nm, and show considerable structural variation. They consist of a thin envelope, a lipid bilayer with tubular or globular surface elements, lens-shaped lateral bodies, and a core. The function of these structures is not always clear. The eight genera of chordopoxviruses, exemplified by the vaccinia virus, have two lateral bodies and surface tubules or grooves. Entomopoxviruses (three genera) are usually occluded in protein crystals and have one or two lateral bodies and concave or biconcave cores.

A. CHORDOPOXVIRUSES

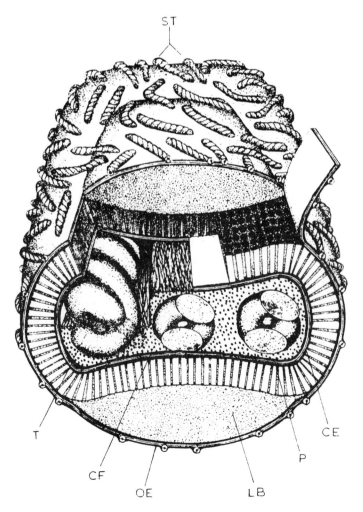

FIGURE 121

Cross-section of vaccinia virus. CE, core envelope; CF, core fibrils; LB, lateral bodies; OE, outer envelope; P, palisade layer; ST, surface tubules; T, triplet elements (nucleoprotein). (From Müller, G. and Williamson, J.D., *Animal Virus Structure*, Nermut, M.V. and Steven, A.C., Eds., Elsevier, Amsterdam, 1987, 421. With permission.)

POXVIRIDAE

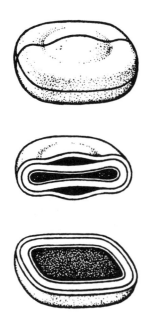

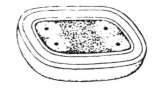

FIGURE 122
Vertical and horizontal sections of vaccinia virus. (Adapted by H.-W. Ackermann from Reference 113, with permission of Nature, © 1956 by Macmillan Magazines Ltd., London.)

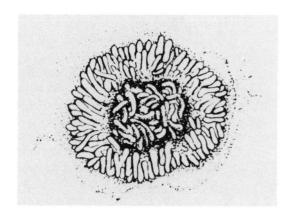

FIGURE 124
Surface view of a member of the *Orthopoxvirus* genus drawn after an electron micrograph. (From Friend Norton, C., *Microbiology*, Addison-Wesley, Reading, MA, 1981, 306. With permission.)

FIGURE 123
Vaccinia "elementary body" in horizontal, vertical, and diagonal sections after OsO$_4$ fixation. (From Peters, D., *Proc. Fourth Int. Congress on Electron Microscopy, Berlin, Sept. 10–17,* Vol. 2, Bargmann, W., Peters, D., and Wolpers, C., Eds. © Springer-Verlag, Berlin, 1958, 552. With permission.)

POXVIRIDAE

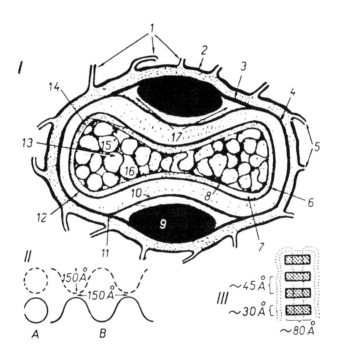

FIGURE 125

Structure of mature variola virus. (I.) Transverse section. 1. "Virovilli." 2 to 4. Outer, intermediate, and internal osmiophilic membranes of envelope. 5. Outer opening of the central hole of a "microvillus." 6 to 8. Outer, intermediate, and internal osmiophilic membranes of nucleoid envelope. 9. Lateral body. 10. Membrane of lateral body. 11. Site of branching of the viral envelope into two sheets, one of which is connected with the membrane of the lateral body. 12. Viroplasm. 13. Nucleoidoplasm. 14. Site of connection of internal osmiophilic membrane of nucleoid envelope with fibrillar component. 15. Osmiophilic fibrils within the cavity of the nucleoid. 16. Hypothetical helices of osmiophilic fibrils. 17. Osmiophilic cavity filling a triangular section. (II.) Hypothetical structure of osmiophilic fibrils in the nucleoid cavity in perpendicular (A) and longitudinal (B) cross-section. (III.) Structure of a part of a "microvillus." (From Avakyan, A.A. and Byrovsky, A.F., *Acta Virol.*, 8, 481, 1964. With permission.)

POXVIRIDAE

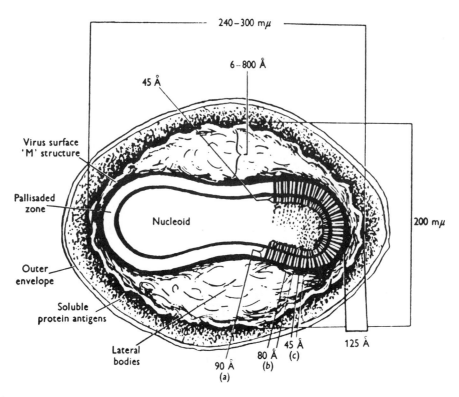

FIGURE 126

Length section of vaccinia virus showing various components and their dimensions. (From Westwood, J.C.N., Harris, W.J., Zwartouw, H.T., Titmuss, D.H.J., and Appleyard, G., *J. Gen. Microbiol.*, 34, 67, 1964. With permission.)

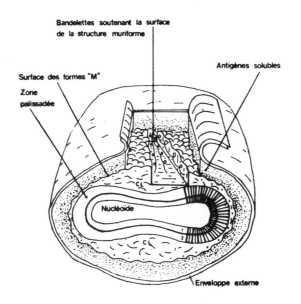

FIGURE 127

Transverse section of a member of the *Poxvirus* genus. The protein-associated viral DNA is located within the nucleoid. The palisade zone, constituted by a regular arrangement of protein subunits, corresponds to the nucleoid membrane. It comprises an internal part of 4.5 nm and a complex external membrane of 9 nm in width. (From Vilaginès, P. and Vilaginès, R., *Virologie médicale*, Maurin, J., Ed., Flammarion, Paris, 1985, 310. With permission.)

POXVIRIDAE

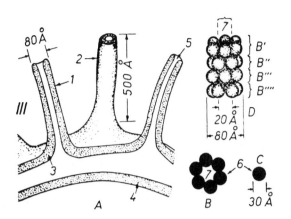

FIGURE 128

Structure of vaccinia virus envelope. A. Section of envelope of mature virus. 1. Section of "virovillus". 2. Intact "virovillus". 3 and 4. Outer and inner osmiophilic membrane of envelope. 5. Central hole of "virovillus". B. Structural unit of "virovillus" composed of 6 subunits and a central hole (6 and 7). C. A single subunit. D. Part of "virovillus" showing arrangement of structural subunits (B', B"). (From Byrovsky, A.F., *Acta Virol.*, 8, 490, 1964. With permission.)

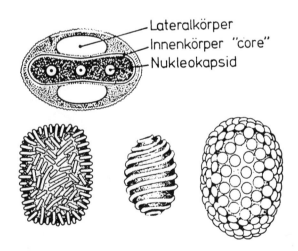

FIGURE 129

Comparative structure of poxviruses. Above: cross-section of orthopoxvirus. Bottom row: surface views of negatively stained orthopox-, parapox-, and entomopoxviruses. Note differences in size and surface structure. (With the kind permission of Blackwell Wissenschafts-Verlag GmbH. Taken from Horzinek, C., *Kompendium der allgemeinen Virologie*, 2nd ed., Paul Parey, Berlin, 1985, 47.) Author's note: parapoxviruses, represented by Orf virus (contagious pustular dermatitis of sheep) are characterized by a single continuous surface thread that covers the whole virus in a criss-cross pattern; entomopoxviruses (right) present a mulberry-like surface.

POXVIRIDAE

FIGURE 130

Three views of surface threads in Orf virus (genus *Parapoxvirus*). The most common pattern is seen at left. (From Nagington, J., Newton, A.A., and Horne, R.W., *Virology*, 23, 461, 1964. With permission.)

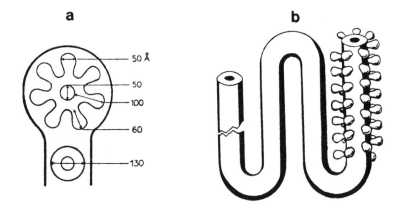

FIGURE 131

Cross-sections of the nucleoid of fowlpox virus (genus *Avipoxvirus*) show a flower-like element (a), believed to be part of a folded tubular structure of 70 to 140 nm in length (b). Its aspect is due to the presence of subunits ("petals") of 5 × 6 nm. (From James, M.H. and Peters, D., *J. Ultrastruct. Res.*, 35, 626, 1971. With permission.)

POXVIRIDAE

B. ENTOMOPOXVIRUSES

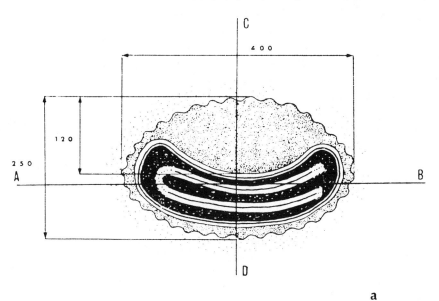

a

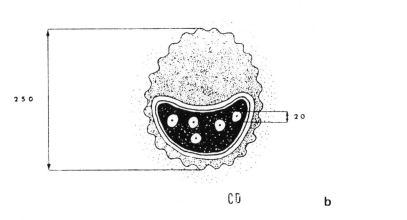

b

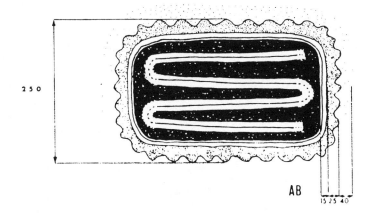

c

FIGURE 132

Melolontha poxvirus in three section planes. Note the asymmetric and eccentric nucleoid. (From Bergoin, M., Devauchelle, G., and Vago, C., *Virology*, 43, 453, 1971. With permission.)

dsDNA, 10–12 segments
Cubic, naked
Vertebrates, invertebrates, plants

XXVII. REOVIRIDAE

This large family has nine genera and a wide host range, encompassing mammals, birds, fish, shellfish, insects, ticks, crustacea, and plants. Members of the genera *Orbivirus* and *Coltivirus* multiply in arthropods and vertebrates. Plant reoviruses multiply in insects and plants. Particles are naked icosahedra of 60 to 80 nm in diameter. Capsids consist of one or two outer coats and an inner core containing 10–12 dsRNA molecules and RNA transcriptase.

Orthoreoviruses have an outer capsid with 120 hexameric and 12 pentameric capsomers (T=13*l*) and a core with 12 apical spikes. Rotaviruses, which cause diarrheas in many mammals, have two outer capsids (T=13*l*) and a core without spikes. Insect reoviruses are grouped in the genus *Cypovirus*, infect insects and crustacea, and are usually occluded in polyhedral protein crystals. Particles are about 60 nm in diameter and have a single shell with hollow apical spikes. Their shells correspond to the cores of other reoviruses. Particles are usually 65 to 70 nm in diameter and possess an outer coat with or without external knobs or spikes. Members of the *Oryzavirus* genus (plants) have no outer coat and consist of a core equivalent with 12 knobs.

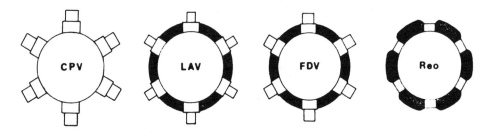

FIGURE 133
Sections of four types of reoviruses. CPV, cytoplasmic polyhedrosis virus (insects); LAV, leafhopper A virus (insects); FDV, Fiji disease virus (plants); Reo, reovirus (genus *Orthoreovirus*, vertebrates). Hatched areas represent the outer capsid shells which can be removed by chymotrypsin (LAV, Reo) or are lost spontaneously (FDV). (From Hatta, T. and Francki, R.I.B., *Intervirology*, 18, 203, 1982. With permission of S. Karger AG, Basel.)

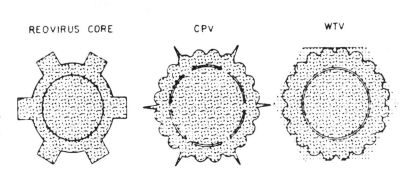

FIGURE 134
Sections of reovirus core, CPV, and wound tumor virus (WTV, plants). (From Lewandowski, L.J. and Traynor, B.L., *J. Virol.*, 10, 1053, 1972. With permission.)

REOVIRIDAE

A. ORTHOREOVIRUSES AND ROTAVIRUSES

Three views of rotaviruses

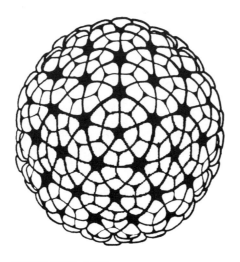

FIGURE 135
Particle with T=9 surface lattice. (From Martin, M.L., Palmer, E.L., and Middleton, P.J., *Virology,* 68, 146, 1975. With permission.)

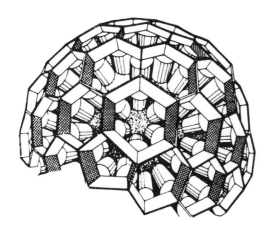

FIGURE 136
Proposed honeycomb structure of the outer capsid and its relationship to the capsomers of the inner capsid. (From Stannard, L.M. and Schoub, B.D., *J. Gen. Virol.,* 37, 435, 1977. With permission.)

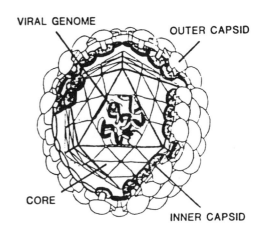

FIGURE 137
View of outer and inner capsid. (Reproduced, with permission from the *Annual Review of Medicine,* Volume 38, p. 399. © 1987, by Annual Reviews Inc., Palo Alto, CA.)

REOVIRIDAE

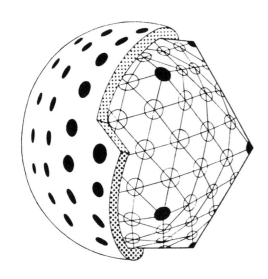

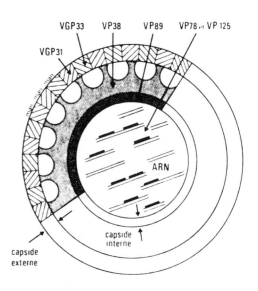

FIGURE 138
Rotavirus particle with 132 coinciding capsomers (T=13) in outer and inner capsid. (From Roseto, A., Escaig, J., Delain, E., Cohen, J., and Scherrer, R., *Virology*, 98, 471, 1979. With permission.)

FIGURE 139
Rotavirus structure. The location of VGP33, VGP31, and VP38 is deduced from concanavalin A treatment and iodination. The position of VP125, VP89, and VP78 is deduced from the study of empty particles and degradation experiments. (From Champsaur, H., *Virologie médicale,* Maurin, J., Ed., Flammarion, Paris, 1985, 716. With permission.)

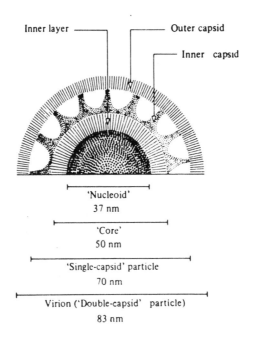

FIGURE 140
Rotavirus structure as deduced from sections of infected cells. (From Esparza, J., Gorziglia, M., Gil, F., and Römer, H., *J. Gen. Virol.*, 47, 461, 1980. With permission.)

REOVIRIDAE

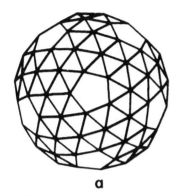

a

FIGURE 141a

Model of reovirus outer capsid (T=13*l*). The complete sphere has 120 triangular facets, 120 vertices, and 12 pentamers. (From Metcalf, P., *Animal Virus Structure*, Nermut, M.V. and Steven, A.C., Eds., Elsevier, Amsterdam, 1987, 135. With permission.)

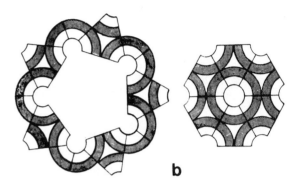

b

FIGURE 141b

Arrangement of triangular capsomers and pentamer crater (left) and hexamer ring according to freeze-etching. Shaded parts represent ridges of the capsid surface. (From Metcalf, P., *Animal Virus Structure*, Nermut, M.V. and Steven, A.C., Eds., Elsevier, Amsterdam, 1987, 135. With permission.)

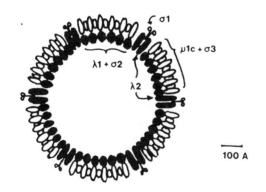

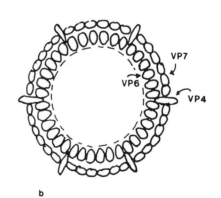

FIGURE 142

Comparison of reoviruses (a) and rotaviruses (b). λ1, core protein; λ2, core spike; μlC and σ3, major outer capsid proteins; σ1, hemagglutinin; σ2, core protein. Core proteins of the former are shown in black. (From Harrison, S.C., *Virology*, 2nd ed., Vol. 1, Fields, B.N. and Knipe, B.M., Eds., Raven Press, New York, 1990, 37. With permission.)

REOVIRIDAE

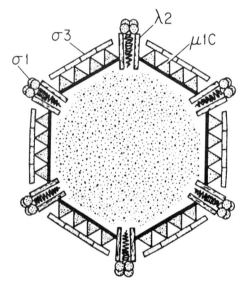

FIGURE 143

Outer capsid of reovirus; see Figure 142 for explanations. (From Schiff, L.A. and Fields, B.N., *Virology,* 2nd ed., Vol. 2, Fields, B.N. and Knipe, D.M., Eds., Raven Press, New York, 1990, 1275. With permission.)

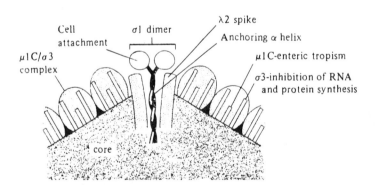

FIGURE 144

Apex of reovirus capsid showing the location of polypeptides with major roles in virulence. The apical spike consists of a globular σ1 dimer at the surface, which is responsible for hemagglutination and attachment, and an α-helical region which anchors σ1 to the λ2 spike protein. The capsid surface consists of a complex of proteins μ1C and σ3. See Figure 142. (From Fenner, F., et al., *Veterinary Virology,* 1st ed., Academic Press, Orlando, FL, 1967, with permission; modified with permission from *Nature,* 315, 421. © 1985 Macmillan Magazines Ltd., London.)

REOVIRIDAE

B. INSECT REOVIRUSES

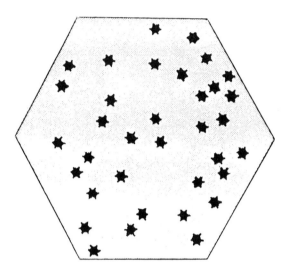

FIGURE 145
Polyhedron with viruses visible at the surface, a common feature in insect reoviruses. (From Nienhaus, T., *Viren, Mykoplasmen und Rickettsien*, Eugen Ulmer, Stuttgart, 1985, 48. With permission.)

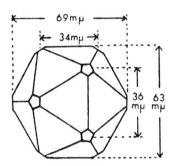

FIGURE 146
Cypovirus dimensions without spikes. (From Hosaka, Y. and Aizawa, K., *J. Insect Pathol.*, 6, 53, 1964. With permission.)

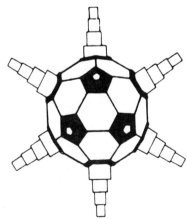 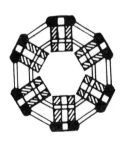

FIGURE 147
A cypovirus compared to a reovirus core. (From Hosaka, Y. and Aizawa, K., *J. Insect Pathol.*, 6, 53, 1964. With permission.)

REOVIRIDAE

C. PLANT REOVIRUSES

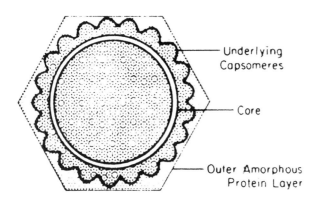

FIGURE 148

Intact wild-type wound tumor virus (WTV) with amorphous outer protein layer (131 and 96 K proteins), underlying capsomers (36 and 335 K), visible after chymotrypsin treatment, and the capsomer-less core which can be seen after CsCl disruption. (From Reddy, D.V.R. and MacLeod, R., *Virology*, 70, 274, 1976. With permission.)

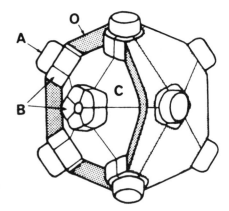

FIGURE 149

Fiji disease virus (FDV) with outer shell (O) partly removed to show the core (C) and structure and arrangement of outer (B) and core spikes (A). (From Hatta, T. and Francki, R.I.B., *Virology*, 76, 797, 1977. With permission.)

Legend of Figure 150: a. Intracisternal particles of type A: resemble cores of types B and D; electron-translucent center, often found in cells producing B and D type particles, biological effects unknown. b. (i) Intracytoplasmic particles of type A: complete cores of B and D type particles, assembled in cytoplasm before budding. (ii) Type B particles: cores assemble in cytoplasm (as type A) and acquire their envelope at the plasma membrane. Cores are eccentric. Spikes are readily visible. c. Type C particles, with central core and poorly visible surface projections. Cores assemble during the budding process at the plasma membrane. d. Type D particles. As type B; cores are assembled in the cytoplasm and centrally located, may assume a tubular shape. Surface projections are as in type C. e. Lentiviruses. Cores assemble at the plasma membrane during the budding process. In contrast to types B, C, and D, core and envelope of budding particles are in close contact. Cores are central and often tubular. Surface projections are knob-like and often poorly visible. f. Spumaviruses. Cores assemble in the cytoplasm. Surface spikes are relatively long and easily visible. Author's note: The subfamilies *Oncornavirinae* (essentially oncogenic viruses), *Lentivirinae* (slow viruses), and *Spumavirinae* (apathogenic "foamy" viruses producing vacuoles in infected cells) are no longer valid. Retroviruses are currently subdivided into seven genera: *Lentivirus, Spumavirus,* and five genera with vernacular names (mammalian type B, C, and D viruses; avian type C viruses, and HTLV/BLV viruses).[12]

ssRNA, 2 segments
Cubic, enveloped
Vertebrates

XXVIII. RETROVIRIDAE

This is a large family with seven genera whose members infect mammals, birds, and reptiles and is frequently associated with tumors, leukemias, anemias, or immunodeficiencies. Particles are spherical, of 80 to 100 nm in diameter, and morphologically diversified. They consist of an envelope provided with glycoprotein projections, an icosahedral or spherical shell, and a central nucleoid or core containing two ssRNA molecules and reverse transcriptase.

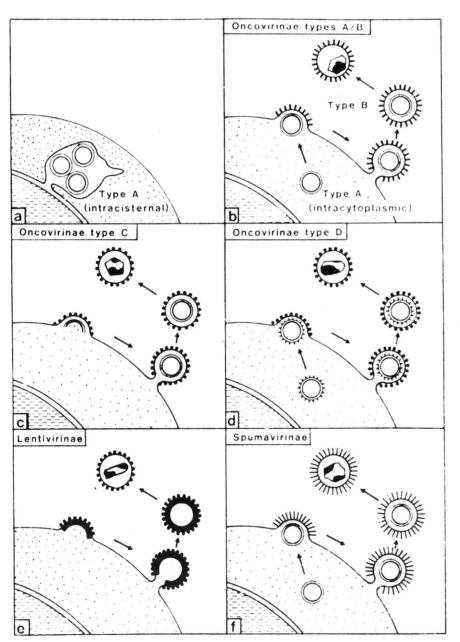

FIGURE 150

Classification of retroviruses by morphology and site of assembly. Legends are on next side. (From Frank, H., *Animal Virus Structure,* Nermut, M.V. and Steven, A.C., Eds., Elsevier, Amsterdam, 1987, 253. With permission.)

RETROVIRIDAE

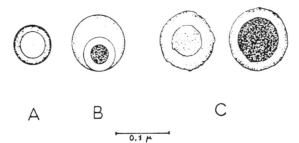

A   B   C

FIGURE 151

Early representation (1960) of virus particles of type A, B, and C retroviruses associated with many mouse tumors and leukemias. Type A, "doughnuts" of unknown significance. Type B probably represents the Bittner mouse mammary tumor virus. Type C is mainly found in leukemias. (From Bernhard, W., *Cancer Res.*, 20, 712, 1960. With permission.)

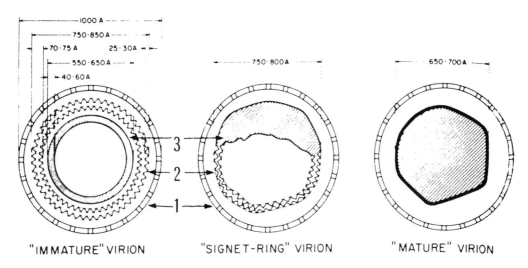

"IMMATURE" VIRION   "SIGNET-RING" VIRION   "MATURE" VIRION

FIGURE 152

Size and structure of different types of particles seen in Rauscher murine leukemia virus (RLV) preparations. The "immature" virion presents (1) an envelope, (2) an outer shell of 7.5 nm in width, corresponding to a hollow, coiled cylinder, and (3) an inner shell that presumably contains the viral RNA. In the "signet-ring" virus, the outer shell appears collapsed over the dense nucleoid. In the "mature" virus, both internal shells are fused together to produce a relatively dense, small nucleoid. (From De-Thé, G. and O'Connor, T.E., *Virology*, 28, 713, 1966. With permission.) Author's note: this relatively outdated figure has a historical value in illustrating stages in retrovirus morphogenesis.

RETROVIRIDAE

Comparison of C-type and lentiviruses

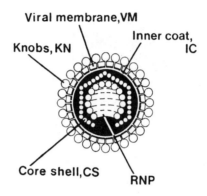

FIGURE 153
C type virus. (From Frank, H. and coll., *Z. Naturforsch.*, 33c, 124, 1978, modified by H. Frank. With permission.)

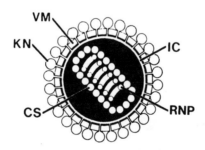

FIGURE 154
Lentivirus. CS, core shell; IC, inner coat; KN, knob; RNP, ribonucleoprotein; VM, viral membrane. (From Frank, H., *Animal Virus Structure*, Nermut, M.V. and Steven, A.C., Eds., Elsevier, Amsterdam, 1987, 295. With permission.)

RETROVIRIDAE

Cutaway diagrams of two generalized tumor retroviruses

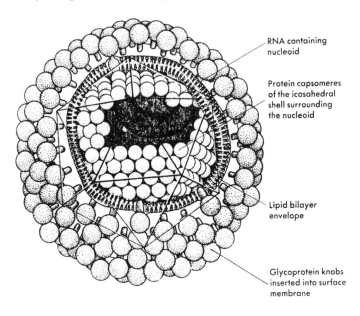

RNA containing
nucleoid

Protein capsomeres
of the icosahedral
shell surrounding
the nucleoid

Lipid bilayer
envelope

Glycoprotein knobs
inserted into surface
membrane

FIGURE 155
(From Watson, J.D., *Molecular Biology of the Gene,* Third
Edition, p. 673. (©) 1976 James D. Watson. Published by
W.A. Benjamin, Inc. With permission.)

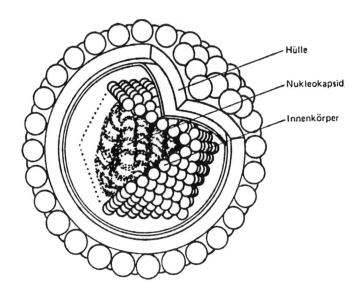

Hülle

Nukleokapsid

Innenkörper

FIGURE 156
(With the kind permission of Blackwell Wissenschafts-Verlag
GmbH. Taken from Horzinek, C., *Kompendium der
allgemeinen Virologie,* 2nd ed., Paul Parey, Berlin,1985, 67.)

RETROVIRIDAE

A. TYPE B VIRUSES

Surface spikes are prominent and cores are eccentric and condensed. Viruses are associated with mammary carcinomas and T-cell lymphomas in mice.

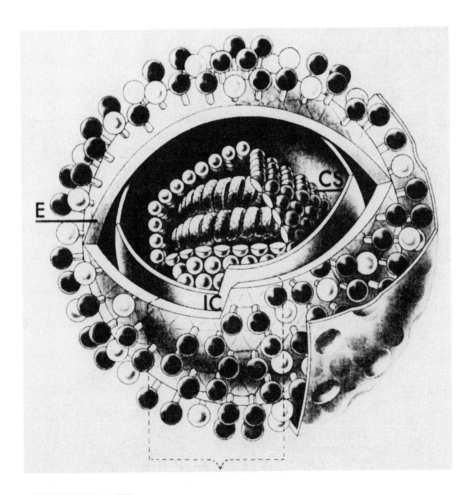

FIGURE 157

Murine mammary tumor virus (MuMTV). At the center of the particle is a core consisting of an RNA/protein complex represented as a helix, and a core shell (CS) with possibly icosahedral symmetry. The envelope (E) is covered with hexagonal arrays of glycoprotein spikes that may form rings (bracket). Beneath the envelope lies a thin, spherical inner coat (IC). The surface of the particle (lower right) shows pits observed after freeze-dry shadowing. (From Sarkar, N.H., *Animal Virus Structure*, Nermut, M.V. and Steven, A.C., Eds., Elsevier, Amsterdam, 1987, 257. With permission.)

RETROVIRIDAE

Mouse mammary tumor viruses

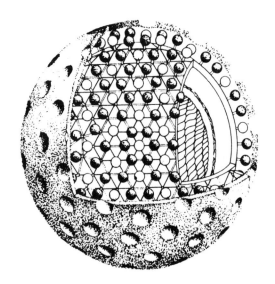

FIGURE 158

Particle showing helical nucleocapsid surrounded by a core shell, a reticular surface structure consisting of hexagons and a few pentagons, two types of projections, and a carbohydrate (?) surface layer with regular holes or pits. (From Sarkar, N.H. and Moore, D.H., *Virology*, 61, 38, 1974. With permission.)

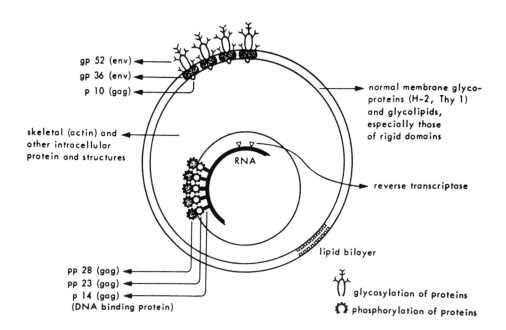

FIGURE 159

Localization of virus proteins. (From Bentvelzen, P. and Hilgers, J., *Viral Oncology*, Klein, G., Ed., Raven Press, New York, 1980, 311. With permission.)

RETROVIRIDAE

B. TYPE C VIRUSES

Surface projections are poorly visible. The capsid is icosahedral (T=25?) and apparently provided with an apical opening.[139] The core is centrally located. Members of the HTLV-BLV group and the *Spumavirus* genus have a similar morphology.

Sections of C type viruses from murine leukemias

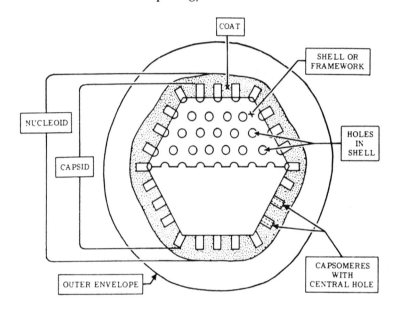

FIGURE 160

Rauscher virus. Capsomers, whose total number has not been determined, are shown for diagrammatical purposes only. (From Padgett, F. and Levine, A.S., *Virology*, 30, 623, 1966. With permission.)

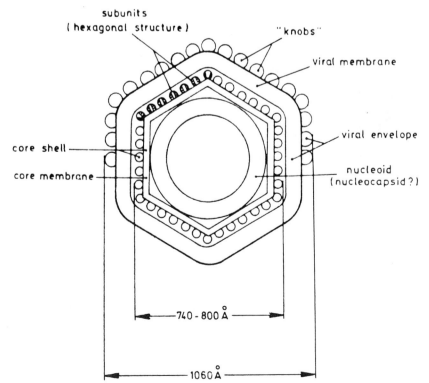

FIGURE 161

Virus structure deduced from conventional techniques, freeze-drying, and freeze-etching. (From Nermut, M.V., Frank, H., and Schäfer, W., *Virology*, 49, 345, 1972. With permission.)

RETROVIRIDAE

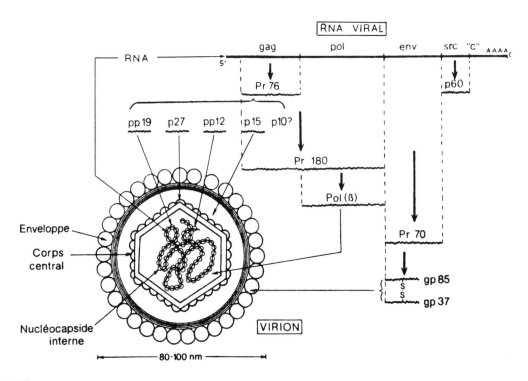

FIGURE 162

Genomic map and structural proteins of avian retroviruses. Most proteins originate by cleavage of precursors. Numerals indicate molecular weights in kilodaltons (K). Proteins: gp, glycoprotein; p, protein; pp, phosphoprotein; Pr, polyproteinic precursor. Genes: *env*, envelope; *gag*, nucleoid; *pol*, reverse transcriptase (about 70 molecules). (From Girard, M. and Hirth, L., *Virologie générale et moléculaire*, Doin, Paris, 1980, 352. With permission.)

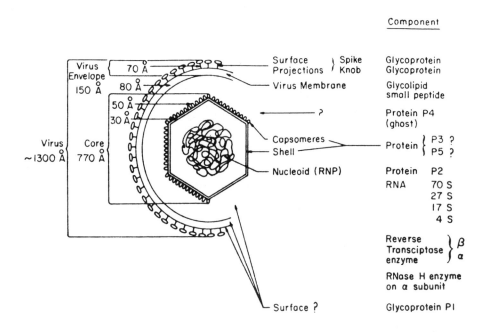

FIGURE 163

Avian myeloblastosis virus with location and nature of components. (From Bolognesi, D.P., *Adv. Virus Res.*, 19, 315, 1974. With permission.)

RETROVIRIDAE

C. LENTIVIRUSES

Cores are rod shaped or resemble truncated cones. Viruses cause immunodeficiencies (e.g., AIDS) or anemias.

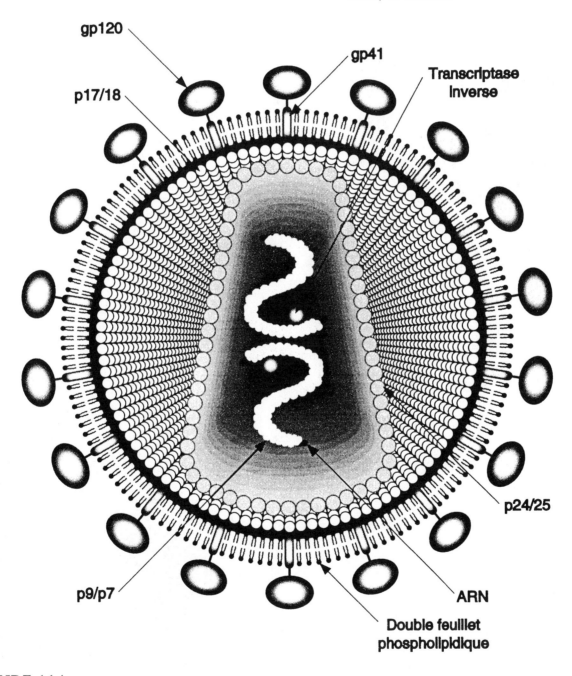

gp120

gp41

Transcriptase Inverse

p17/18

p24/25

p9/p7

ARN

Double feuillet phospholipidique

FIGURE 164

The AIDS virus (HIV-1). Knobs made of protein gp120 are anchored to transmembrane protein gp41. The wedge-shaped core shell consists of a protein called p24 or p25 and contains RNA and reverse transcriptase. (By G. Larose, modified from Reference 150, courtesy of L. Thibodeau.)

RETROVIRIDAE

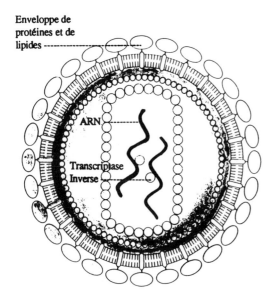

Enveloppe de
protéines et de
lipides

ARN

Transcriptase
Inverse

FIGURE 165
A different version with a tubular nucleoid. Original in color. (From Gallo, R.C. and Montagnier, L., AIDS in 1988, *Sci. Am.*, 259(4), 41, 1988. With permission.)

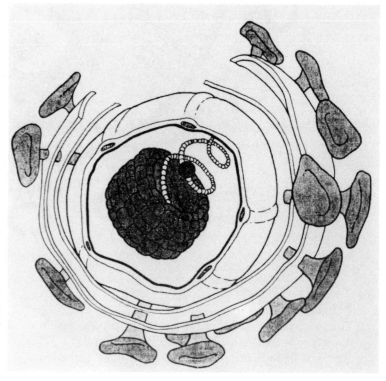

FIGURE 166
AIDS virus with location of components. The diagram, originally in color, was part of a larger drawing of which some elements intruded into the virus particle. Original in color. (From Boyd, J.E. and James, K., *Microbiol. Sci.*, 5, 300, 1988. With permission.)

RETROVIRIDAE

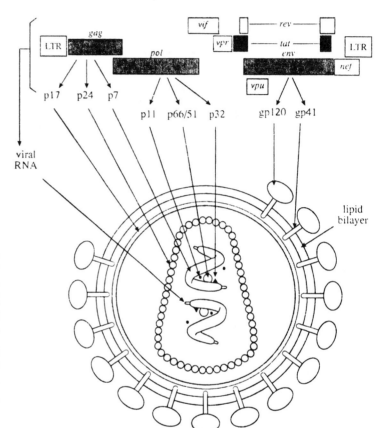

FIGURE 167

Genomic map and components of AIDS virus. Principal genes are *gag*, *pol*, and *env*; they code for polyproteins that are processed by proteolytic cleavage. The *gag* gene codes for membrane-associated protein p17, the major capsid protein p24, and nucleocapsid protein p7. The *pol* gene codes for reverse transcriptase, protease, and integrase (RT, PR, IN) enzymes. Surface and transmembrane proteins (SU, TM) are specified by the *env* gene. (From Bolognesi, D.P., *Adv. Virus Res.*, 42, 103, 1993. With permission.)

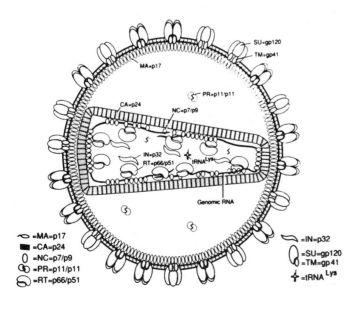

FIGURE 168

AIDS virus with location of components. Abbreviations are as in Figure 167. (From Arnold, E. and Arnold, G.F., *Adv. Virus Res.*, 39, 1, 1991. With permission.)

RETROVIRIDAE

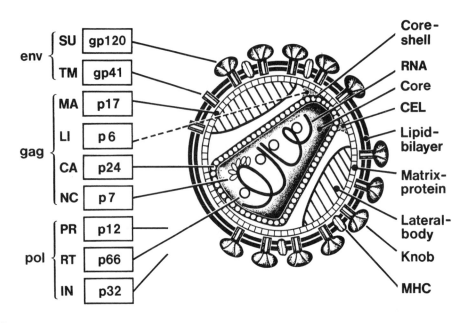

FIGURE 169

Structure and components of AIDS virus. The number of capsomers of the core shell is still unknown. Particles contain ill-defined lateral bodies. The number of surface "knobs" is estimated at 70 to 80. CA, capsid; CEL, core-envelope link; IN, integrase; LI, link; MA, membrane-associated; MHC, mouse-specific histocompatibility antigen; NC, nucleocapsid; PR, protease; RT, reverse transcriptase; SU, surface; TM, transmembrane. (From Gelderblom, H.R., Gentile, M., Scheidler, A., Özel, M., and Pauli, G., *AIDS Forsch.*, 8, 231, 1993. With permission.)

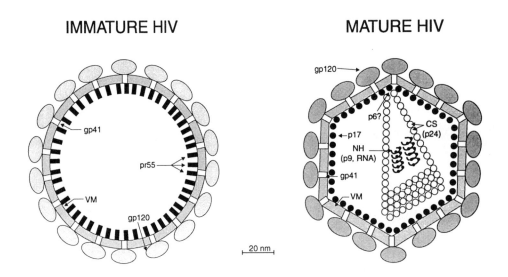

FIGURE 170

Immature and mature AIDS virus according to present knowledge of gag proteins. During maturation, the pr55 protein is cleaved and its derivative p17 forms an icosahedral shell. Particle diameter is about 120 nm. Dimensions of protein molecules and structural elements: gp120 knobs, 15 × 9 nm; p17, 3.4 nm; p24, 3.8 nm; pr55, 3.4 × 8.5 nm; lipid bilayer, 4 to 5 nm. CS, core shell; NH, nucleohelix; VM, viral membrane. (From Nermut, M.V., Hockley, D.J., Jowett, J.B.M., Jones, I.M., Garreau, M., and Thomas, D., *Virology*, 198, 288, 1994. With permission.)

dsDNA, linear
Cubic, naked
Bacteria

XXIX. TECTIVIRIDAE

This small virus family has 14 known members, multiplying either in bacilli or in Gram-negative bacteria harboring certain plasmids (enterobacteria and their relatives). Particles are naked icosahedra of about 63 nm in diameter. Capsids consist of an outer thin, rigid, proteinaceous shell and an inner, thick-walled lipoprotein vesicle. Upon adsorption to bacteria or chloroform treatment, the vesicle becomes a tail-like tube of 60 × 10 nm. *Bacillus* tectiviruses have 20-nm long fibers on vertices.

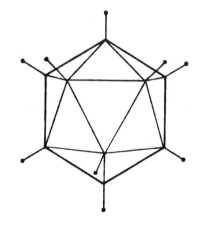
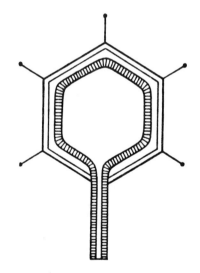

FIGURE 171

Bacillus phage Bam35 in surface view and section, showing apical fibers, a rigid outer protein coat, and a tail-like tube derived from the inner lipoprotein vesicle. (Reprinted from Ackermann, H.-W. and DuBow, M.S., *Viruses of Prokaryotes,* Vol. 2, 1987, 174, CRC Press, Boca Raton, FL.)

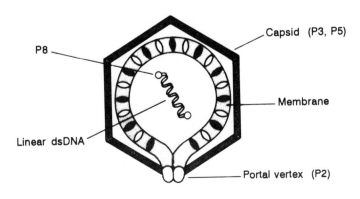

FIGURE 172

Phage PRD1 of enterobacteria and pseudomonads, showing portal vertex and DNA-associated protein. The lipid membrane contains 15 viral proteins. Tectiviruses of Gram-negative bacteria have no fibers. (By H.-W. Ackermann and M. Côté after Reference 158 and unpublished data, with permission of D.H. Bamford.)

ssRNA, + sense
Cubic, naked
Invertebrates

XXX. TETRAVIRIDAE

This small virus family is found in Lepidopteran insects. Particles are naked icosahedra (T=4) of about 40 nm in diameter. The group apparently includes viruses with monopartite and bipartite RNA.

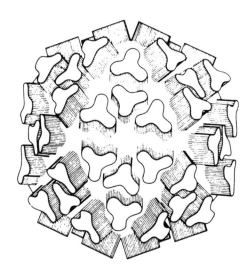

FIGURE 173
Surface morphology of *Nudaurelia capensis* ß virus. (From Finch, J.T., Crowther, R.A., Hendry, D.A., and Struthers, J.K., *J. Gen. Virol.*, 24, 191, 1974. With permission.)

ssRNA, + sense
Helical, naked
Plants

XXXI. GENUS TOBAMOVIRUS

This group comprises the tobacco mosaic virus, one of the first viruses whose structure and assembly were known in detail. Particles are rigid rods of 300×10 nm provided with a central channel. RNA and coat protein form an intertwined helix.

First detailed descriptions (1956 and 1957) of tobacco mosaic virus (TMV)

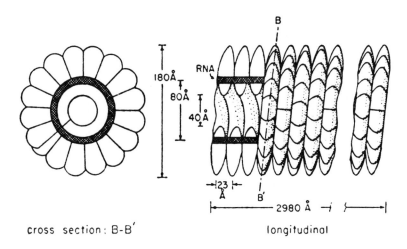

cross section: B-B′   longitudinal

FIGURE 174

Cross-section and longitudinal view. The RNA is thought to be a cylinder. (From Stanley, W.M., *Fed. Proc.*, 15, 812, 1956. With permission.)

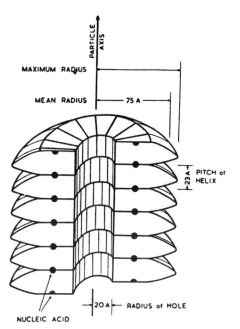

FIGURE 175

Length section of a small part of the virus, showing helical arrangement of protein subunits (49 per 3 turns of helix). The RNA is now represented as a spiral. (From Franklin, R.E., Klug, A., and Holmes, K.C., *The Nature of Viruses*, Wolstenholme, G.E.W. and Millar, E.C.P., Eds., Churchill Livingstone, London, 1957, 39. With permission.)

TOBAMOVIRUS

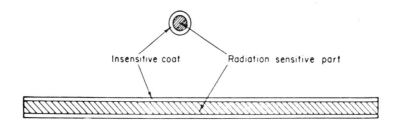

FIGURE 176

Structure of TMV as deduced from radiation studies. (From Pollard, E.C., *Hepatitis Frontiers,* p. 355, Hartman, F.W., LoGrippo, G.A., Mateer, J.G., and Barron, J., Eds. © 1957 Little, Brown and Company, Boston. With permission.)

FIGURE 177

A famous diagram of TMV (1960) that has found its way into many textbooks of microbiology. For clarity, part of the RNA is shown without supporting protein. Each protein subunit corresponds to 3 nucleotides (49 per turn of helix), represented by small disks in accordance with the predominant orientation of purine and pyrimidine bases. (From Klug, A. and Caspar, D.L.D., *Adv. Virus Res.,* 7, 225, 1960. With permission.)

TOBAMOVIRUS

DLDC

FIGURE 178

Structure of a 15-turn segment of dried TMV, as inferred from electron microscopy, and the structure of the virus in solution. Because of local variations in drying forces, it is probable that the actual clustering of subunits is more disordered than indicated. (From Caspar, D.L.D., *Plant Virology*, Corbett, M.K. and Sisler, H.D., Eds., University of Florida Press, Gainesville, FL, 1964, 267. With permission.)

TOBAMOVIRUS

Comparison of TMV and other filamentous plant viruses

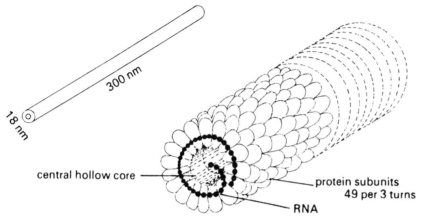

central hollow core

protein subunits
49 per 3 turns

RNA

Figure 179

FIGURES 179 AND 180
Rigid TMV (179) and flexible potatovirus X (genus *Potexvirus*, 180). (From Stevens, W.A., *Virology of Flowering Plants*, Blackie & Sons, Glasgow, 1983, 79 and 80. With permission.)

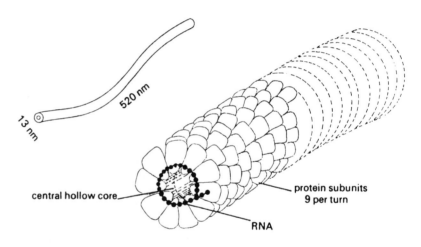

central hollow core

protein subunits
9 per turn

RNA

Figure 180

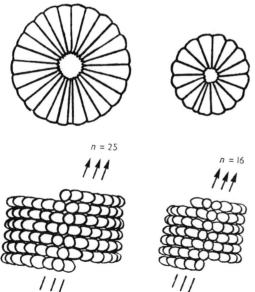

$n = 25$

$n = 16$

(a)   (b)

FIGURE 181
Scale drawings in plan and elevation of (a) tobacco rattle virus (genus *Tobravirus*) and (b) TMV; n, approximate number of subunits per turn of helix. (From Offord, R.E., *J. Mol. Biol.*, 17, 370, 1966. With permission.)

ssRNA, + sense
Cubic, enveloped
Vertebrates, invertebrates

XXXII. TOGAVIRIDAE

This small family includes the genera *Alphavirus* (type A arboviruses) and *Rubivirus* (rubella virus). Particles are spherical, 60 to 70 nm in diameter, and consist of an envelope and a capsid. In alphaviruses, the tight-fitting envelope is studded with glycoprotein spikes that are arranged as trimers in a T=4 icosahedral lattice. The icosahedral capsid (T=4) is fenestrated. Rubiviruses are pleomorphic and their structure is not known in detail.

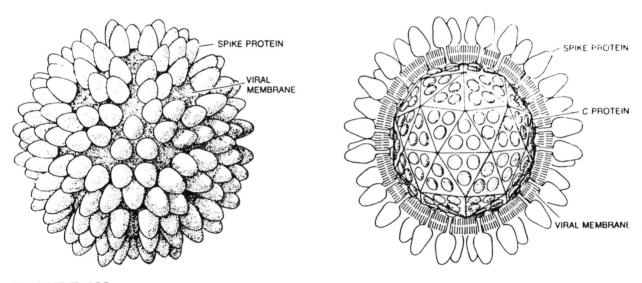

FIGURE 182

Structure of Sindbis virus (genus *Alphavirus)*. The envelope is a lipid bilayer with the hydrophobic side outward and the hydrophilic side inward. It has 180 "spikes", each consisting of 3 linked protein molecules. Each spike is bound to a molecule of capsid protein C. The capsid has 60 faces, each of which is an equilateral triangle with 3 molecules of C protein. (From Simons, K., Garoff, H., and Helenius, A., *Sci. Am.,* 246(2), 58, 1982. © Scientific American, Inc., New York. All rights reserved.)

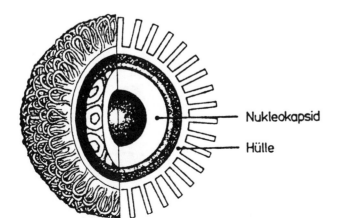

FIGURE 183

Cutaway diagram of an alphavirus. (With the kind permission of Blackwell Wissenschafts-Verlag GmbH. Taken from Horzinek, C., *Kompendium der allgemeinen Virologie,* 2nd ed., Paul Parey, Berlin, 1985, 58.)

TOGAVIRIDAE

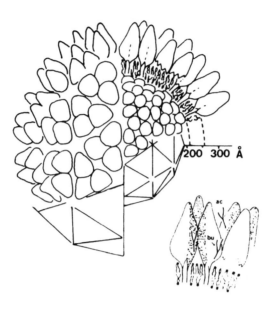

FIGURE 184

Sindbis virus structure. Spikes cluster in trimers. Each spike consists of one E1 and one E2 subunit; both E1 and E2 have hydrophobic anchors that penetrate the envelope. E2 has an internal peptide of 33 amino acid residues that presumably makes contact with the C protein of the capsid. At right, location of buried (bu) and accessible (ac) oligosaccharides. (From Harrison, S.C., *Adv. Virus Res.,* 28, 175, 1983. With permission.)

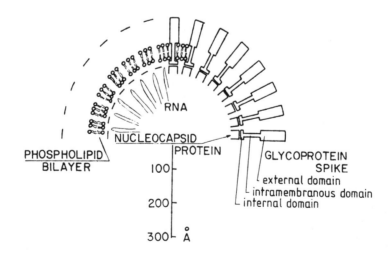

FIGURE 185

Semliki Forest virus structure (genus *Alphavirus*). Glycoprotein spikes span the envelope and connect with the nucleoplasid. (Reprinted from Simons, K., Garoff, H., and Helenius, A., *Membrane Proteins and their Interactions with Lipids,* Vol. 1, Capaldi, R., Ed., 1977, by courtesy of Marcel Dekker, Inc., New York.)

TOGAVIRIDAE

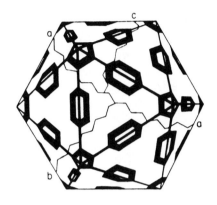

FIGURE 186

Sindbis virus nucleocapsid. The diagram shows a T=4 surface lattice observed along an axis of threefold symmetry. Thin serrate lines indicate domains where drying forces might produce fractures and generate two-, three-, or four-lobed structures. (From Coombs, K. and Brown, D.T., *Virus Res.*, 7, 131, 1987. With permission.)

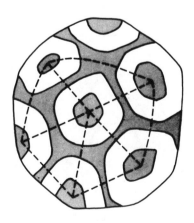

FIGURE 187

Sindbis virus core, showing distribution of electron density and correlation of subunits with a T=3 surface lattice. (By H.-W. Ackermann and M. Côté after a model in Reference 169.)

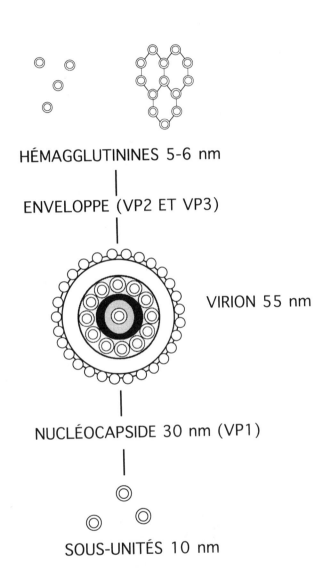

HÉMAGGLUTININES 5-6 nm

ENVELOPPE (VP2 ET VP3)

VIRION 55 nm

NUCLÉOCAPSIDE 30 nm (VP1)

SOUS-UNITÉS 10 nm

FIGURE 188

Structure of rubella virus. The envelope has hemagglutinating subunits able to assemble to hexagonal rings. The nucleocapsid is composed of subunits of 10 nm in diameter. (From Payment, P., et al., *Can. J. Microbiol.*, 21, 703, 1975. With permission.)

ssRNA, +, 1–3 segments
Cubic, naked
Plants

XXXIII. ssRNA PLANT VIRUSES WITH CUBIC SYMMETRY

This large and heterogeneous group comprises nonenveloped viruses of about 30 nm in diameter that belong to several families and not further classified "floating" genera. Particles are icosahedral with generally 32 capsomers (T=3 or T=1). Some viruses constitute multicomponent systems with two or three RNA molecules encapsidated in different particles. Others are morphologically heterogeneous and include various size classes of isometric or more or less elongated particles.

Early and recent representations

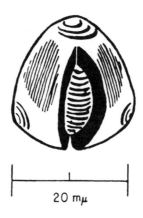

20 mμ

FIGURE 189

Turnip yellow mosaic virus (TYMV, genus *Tymovirus*). In this first diagram of any plant virus, a sector of the coat has been removed to show the relative volume occupied by the nucleic acid. The scale marker refers to a dry particle. (From Markham, R., *Adv. Virus Res.,* 1, 315, 1953. With permission.)

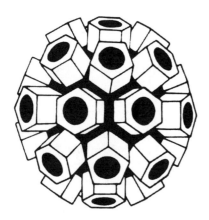

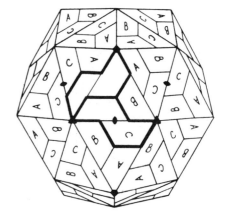

FIGURE 190

Cowpea chlorotic mottle virus (CPMV, family *Bromoviridae*). Capsomers forming a T=3 capsid are viewed along an axis of two-fold symmetry. (From Horne, R.W., Hobart, J.M., and Pasquali-Ronchetti, I., *J. Ultrastruct. Res.,* 53, 319, 1975. With permission.)

FIGURE 191

Southern bean mosaic virus (SBMV, genus *Sobemovirus*) showing distribution pattern of subunits (T=3). (From Rossmann, M.G. and Rueckert, R.R., *Microbiol. Sci.,* 4, 208, 1987. With permission.)

ssRNA PLANT VIRUSES WITH CUBIC SYMMETRY

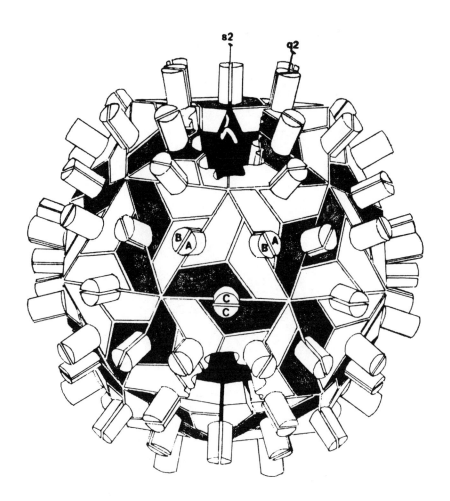

FIGURE 192

Tomato bushy stunt virus (TBSV, family *Tombusviridae*) showing arrangement of subunits (A, B, C) with their packing environments (T=3). Outer surfaces of C subunit S domains are shaded. S domains of A subunits pack around fivefold axes and those of B and C alternate around threefold axes. The shell has been opened in two places to reveal S domain packing. (Reprinted with permission from *Nature*, 276, 368. © 1978 Macmillan Magazines Ltd., London.)

ssRNA PLANT VIRUSES WITH CUBIC SYMMETRY

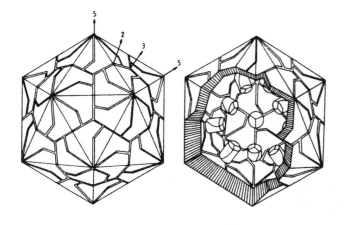

FIGURE 193

Satellite tobacco mosaic virus (STMV, genus *Necrovirus*) showing subunit packing (T=3). (From Liljas, L. and Strandbert, B., *Biological Macromolecules and Assemblies*, Vol. 1, p. 97, Jurnak, F.A. and McPherson, A., Eds., © 1984 by John Wiley & Sons, New York. Reprinted by permission.)

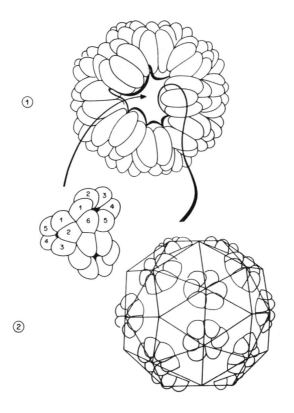

FIGURE 194

TYMV structure. (1) The RNA is closely associated with the subunits of the shell and not simply located in the center of the particle. Two hexamers and one pentamer have been removed from the capsid. (2) After complete removal of pentamers, each triangle of the T=3 net has only two subunits left; the third subunit is thus always part of a pentamer. (From Cornuet, P., *Eléments de virologie végétale*, INRA, Paris, 1987, 37. With permission.)

ssRNA PLANT VIRUSES WITH CUBIC SYMMETRY

Four ways to represent a T=3 capsid

FIGURE 195

TYMV as revealed by negative staining. The 180 subunits protrude about 2 nm from the body of the particle; their exact shape is not known. Subunits are tilted toward the three and fivefold axes of the particle. (From Finch, J.T. and Klug, A., *J. Mol. Biol.*, 15, 344, 1966. With permission.)

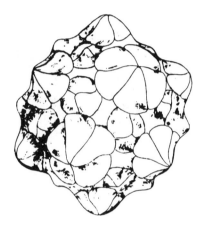

FIGURE 196

CPMV model based on three-dimensional reconstruction and protein composition of the capsid. (From Van Kammen, A. and Mellema, J.E., *The Atlas of Plant and Insect Viruses*, Maramorosch, K., Ed., Academic Press, New York, 1977, 167. With permission.)

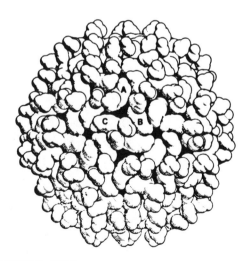

FIGURE 197

TBSV after expansion in the presence of divalent cations. Subunits are represented as consisting of five lumps (three for S domains, two for P). (From Harrison, S.C., *Adv. Virus Res.*, 28, 175, 1983. With permission.)

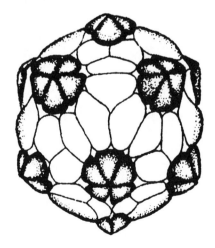

FIGURE 198

Comovirus particle (family *Comoviridae*). Capsids consist of 12 × 5 apical and 20 × 3 lateral subunits. (From Stevens, W.A., *Virology of Flowering Plants*, Blackie & Sons, Glasgow, 1983, 89. With permission.)

ssRNA PLANT VIRUSES WITH CUBIC SYMMETRY

Alfalfa mosaic virus (AMV, family *Bromoviridae*)

FIGURE 199

Five size classes showing derivation of bacilliform particles from a T=1 icosahedron. (With the kind permission of Blackwell Wissenschafts-Verlag GmbH. Taken from Horzinek, C., *Kompendium der allgemeinen Virologie*, 2nd ed., Paul Parey, Berlin, 1985, 72.)

top a (Ta)

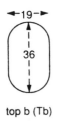

top b (Tb)

middle (M)

bottom (B)

FIGURE 200

Dimensions of the four main classes of particles (nm). (From Matthews, R.E.F., *Plant Virology*, 3rd ed., Academic Press, San Diego, 1991, 127. With permission.)

FIGURE 201

Size classes of Ourmia melon mosaic virus (nonclassified). Superposed exposures of electron micrographs indicate that particles consist of several packages of double or triple disks. (From Matthews, R.E.F., *Plant Virology*, 3rd ed., Academic Press, San Diego, 1991, 129. With permission.)

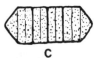
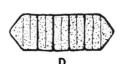
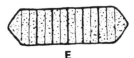

ssRNA PLANT VIRUSES WITH CUBIC SYMMETRY

FIGURE 202

Proposed arrangement of coat protein subunits in AMV. The 30 dimers have been positioned over the lattice of the T=1 icosahedron in a way to coincide with the twofold symmetry axes of the particle. Solid lines indicate bonds between dimers. (From Driedonks, R.A., Krijgsman, P.C.J., and Mellema, J.E., *J. Mol. Biol.*, 113, 123, 1977. With permission.)

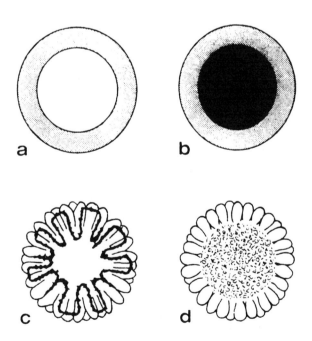

FIGURE 203

Distribution of RNA and protein in TYMV. a and b. Markham's concept of an empty protein shell that is filled with RNA. c. Interpenetration of RNA and protein according to X-ray crystallographic analysis; probably a misinterpretation due to high salt concentration in the medium. d. Distribution of RNA according to low-angle neutron scattering, showing little interpenetration. (From Matthews, R.E.F., *Intervirology*, 15, 121, 1981. With permission of S. Karger AG, Basel.)

XXXIV. MISCELLANY

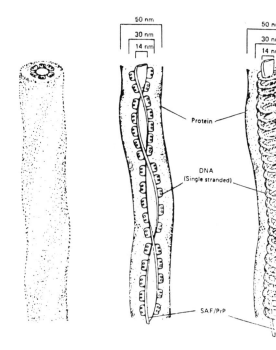

FIGURE 204

Scrapie-associated filaments ("nemaviruses") showing two possible locations for ssDNA. (From Narang, H.K., *Intervirology*, 34, 105, 1992. With permission of S. Karger AG, Basel.)

FIGURE 205

Potato spindle tuber viroid. (From Matthews, R.E.F., *Fundamentals of Plant Virology*, Academic Press, San Diego, 1992, 185. With permission.)

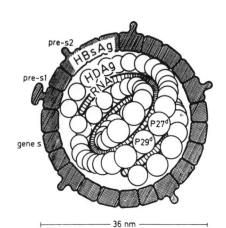

FIGURE 206

Deltavirus of hepatitis D. A defective virus that needs the presence of hepatitis B virus for replication and borrows its envelope from this agent. Hepatitis delta antigen (HDAg) is an internal component of the virus and consists of two proteins (P27, P29). The envelope contains S and pre-S polypeptides of hepatitis B virus. (From Zyzik, E., Ponsetto, A., Forzani, B., Hele, C., Heermann, K.-H., and Gerlich, W.H., *Hepadna Viruses*, Robinson, W., Koike, K., and Will, H., Eds., Alan R. Liss, New York, 1987, 565. With permission.)

Chapter 5
VIRUSES WITH BINARY SYMMETRY

dsDNA, linear
Binary, naked
Bacteria

Particles consist of heads and tails of cubic and helical symmetry, respectively. Viruses of this type, commonly called "tailed phages," are specific to bacteria. With over 4000 isolates studied by electron microscopy and a host range that includes archae-, eu-, and cyanobacteria, they are the largest and probably oldest virus group of all. Tailed phages are extremely diversified with respect to particle size, fine structure, and physicochemical and biological properties.

Capsids are icosahedra or elongated derivatives thereof and measure usually about 60 nm in diameter (range 45 to 170 nm). Elongated heads are up to 230 nm long. Capsomers are seldom visible. Tails are composed of helical rows or stacked disks of subunits and are usually provided with fixation organelles (base plates, spikes, fibers) for adsorption to the bacterial surface. Tailed phages are classified into three families. Each family has only one genus because there are no generally accepted criteria for establishing genera in these viruses.

FIGURE 207
Several types of tailed bacteriophages, then named "Herellen," are seen in this first graphical representation of any virus (probably an ink drawing that reproduces the aspect of unstained electron-dense particles). Although the particles at left and right show no tail, they must be short-tailed podoviruses because the visualization of phages of the φX174 and MS2 types was beyond the reach of early electron microscopes. The "tailless" particles may represent the T7 and 7–11 species of coliphages (see Figure 258). (From Ruska, H., *Ergeb. Hyg. Bakteriol. Immunforsch. Exp. Ther.*, 25, 437, 1943. © Springer-Verlag. With permission.)

Tail contractile

I. MYOVIRIDAE

Heads are isometric or elongated. Tails are long, thick (80 to 485 × 16 to 20 nm), rigid, and contractile, consisting of a central tube or core and a sheath that is separated from the head by a neck. During contraction, the sheath becomes shorter and thicker, the core is exposed, and base plate and tail fibers expand or unfold. The family is typified by T-even phages (T2, T4, T6) and has about 1000 members.

Three very early diagrams

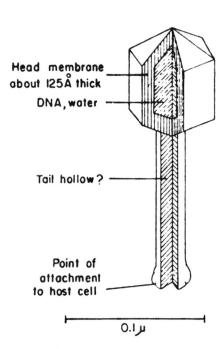

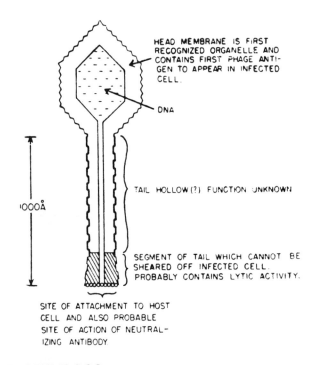

FIGURE 208

Coliphage T2 with a sector of the head cut away (1952). (From Anderson, T.F., et al., *Ann. Inst. Pasteur,* 84, 5, 1953. With permission.)

FIGURE 209

A T-even type phage (1953). (From Anderson, T.F., *Cold Spring Harbor Symp. Quant. Biol.,* 18, 197, 1953. With permission.)

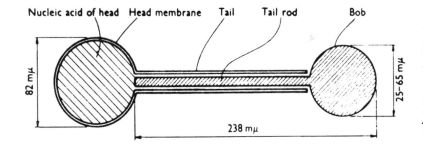

FIGURE 210

Staphylococcal phage K (1954). The phage was only in 1981 recognized to have a contractile tail. The "bob" at the end of the tail must be a cell wall debris.[188] (From Hotchin, J.E., *J. Gen. Microbiol.,* 10, 250, 1954. With permission.)

MYOVIRIDAE

Two early diagrams of coliphage T2

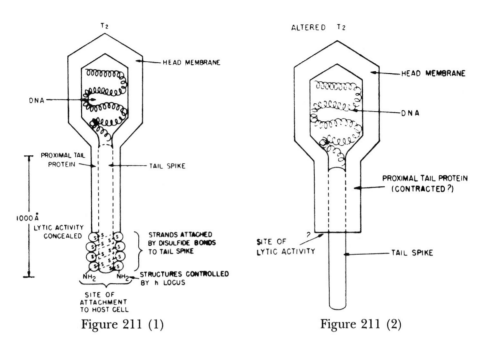

Figure 211 (1)   Figure 211 (2)

FIGURE 211
Phage with (1) extended and (2) contracted tail (1956). (From Evans, E.A., *Fed. Proc.*, 15, 827, 1956. With permission.)

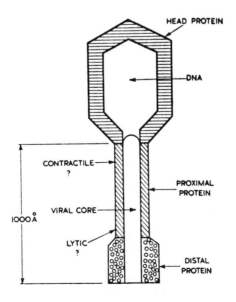

FIGURE 212
Particle with extended tail (1959). (From Evans, E.A., *The Viruses*, Vol. 1, Burnet, F.M. and Stanley, W.M., Eds., Academic Press, New York, 1959, 459. With permission.)

MYOVIRIDAE

Two diagrams of T-even phages by the same author

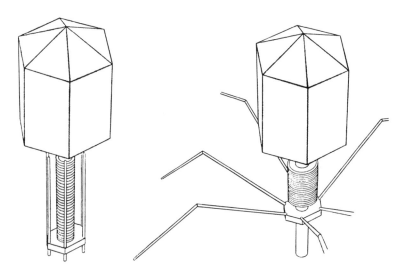

FIGURE 213
Phage T2 before and after tail contraction (1963). The head is represented as a bipyramidal hexagonal prism. In quiescent phages, the six tail fibers are folded along the tail. Some tail fibers of the right particle are incomplete and the number of tail striations is too high (author's note). (From Horne, R.W., *Sci. Am.*, 147, 2, 1963. © Scientific American, Inc., New York. All rights reserved.)

FIGURE 214
Phage with extended tail and tail fibers (1975). The head is now represented as a prolate icosahedron. The tail sheath shows subunits and terminates in a hexagonal base plate. (From Horne, R.W., *Adv. Optical Electron Microsc.*, 6, 227, 1975. With permission.)

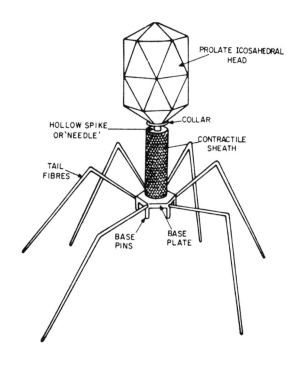

MYOVIRIDAE

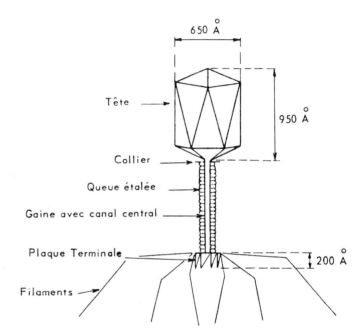

650 Å

950 Å

Tête

Collier

Queue étalée

Gaine avec canal central

Plaque Terminale

Filaments

200 Å

FIGURE 215

Phage T2 with extended tail and tail fibers. The head is shown as a bipyramidal pentagonal antiprism, an incomplete form of a prolate icosahedron and the true shape of the phage head. (From Moustardier, G., *Virologie médicale,* 4th ed., Maloine, Paris, 1973, 138. With permission.)

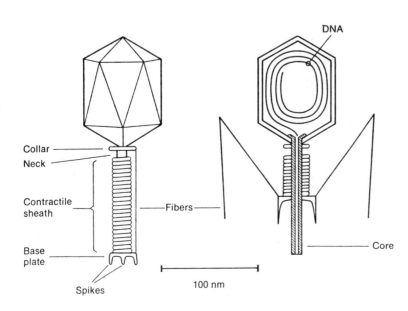

DNA

Collar

Neck

Contractile sheath

Fibers

Base plate

Spikes

Core

100 nm

FIGURE 216

Phage T4 with extended and contracted tail (24 and 12 transverse striations, respectively). (From Ackermann, H.-W. and DuBow, M.S., *Viruses of Prokaryotes,* Vol. 2, 1987, 2, CRC Press, Boca Raton, FL.)

MYOVIRIDAE

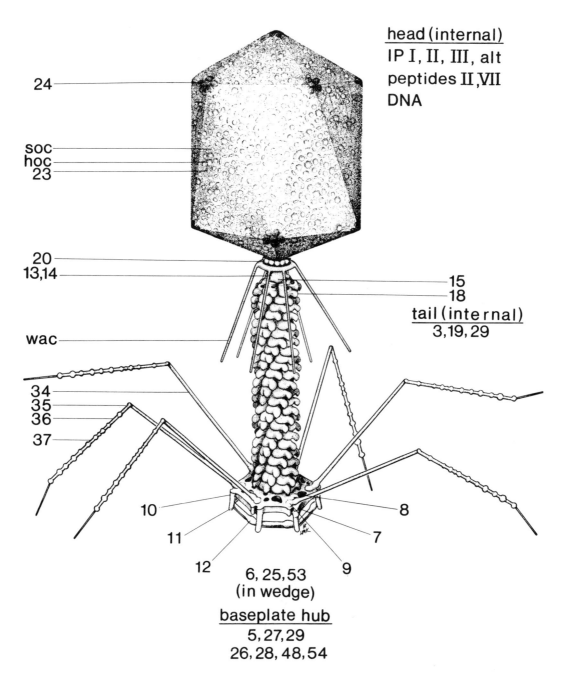

24

soc
hoc
23

20
13,14

wac

34
35
36

37

10

11

12

head (internal)
IP I, II, III, alt
peptides II,VII
DNA

15
18
tail (internal)
3,19,29

8

7

9

6, 25, 53
(in wedge)
baseplate hub
5, 27, 29
26, 28, 48, 54

FIGURE 217
A splendid model of phage T4 (1983) showing detailed location of structural proteins. Head vertices consist of cleaved gp24. Gp20 is located at the head-tail connector. Collar and whiskers appear to be made of the same protein, gp*wac*. Sheath subunits (gp18) fit into holes in the base plate and short tail fibers (gp12) are in the quiescent state. The very complex base plate is assembled from a central plug and six wedges. Tail fibers consist of three proteins. (From Eiserling, F.A., *Bacteriophage T4*, Mathews, C.K., Kutter, E.M., Mosig, G., and Berget, P.B., Eds., American Society for Microbiology, Washington, DC, 1983, 11. Revised with permission.)

MYOVIRIDAE

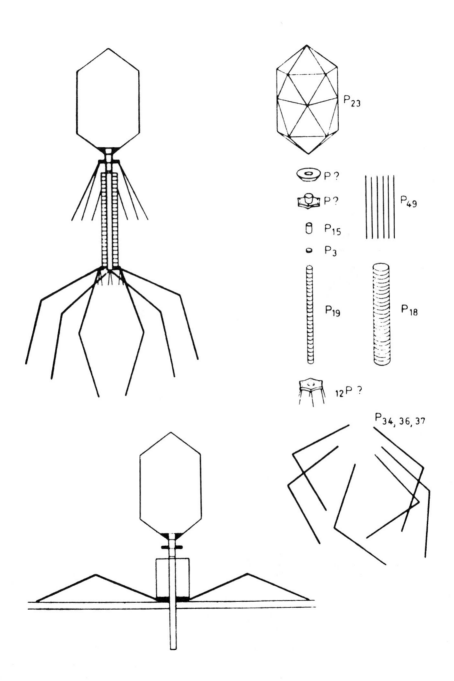

FIGURE 218

Structure and components of T-even phages. The particle below is in the process of infecting a bacterium. (From Poglazov, B.F., *Monographs in Developmental Biology*, Vol. 7, S. Karger AG, Basel, 1973, 7. With permission.)

MYOVIRIDAE

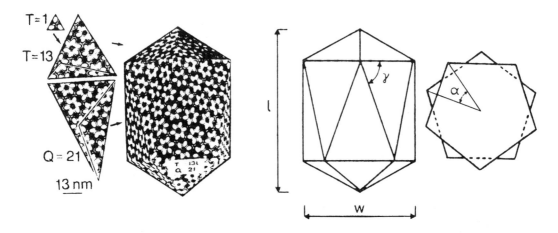

FIGURE 219

Geometry of T-even phage heads. At left, folding of the basic hexagonal lattice into a prolate icosahedron. Caps consist of five equilateral triangles each. The central part of the phage head consists of ten elongated triangles. The model shown is a T=13l and Q=21 prolate icosahedron. At right, capsid parameters specifying the Q number for a given T number: length and width *(l, w)* of the prolate icosahedron, facet angle g specifying elongated triangles, eclipse angle a specifying relative rotation of caps. (From Baschong, W., Aebi, U., Baschong-Prescianotto, C., Dubochet, J., Landmann, L., Kellenberger, E., and Wurtz, M., *J. Ultrastruct. Res.,* 99, 189, 1988. With permission.)

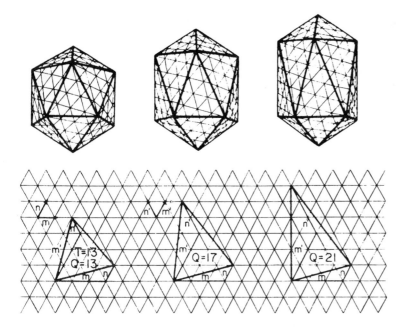

FIGURE 220

Surface lattices predicted for isometric, intermediate, and normal T4 heads (above) and definitions of lattice parameters (below). In isometric heads, the surface lattice is defined by the triangulation number T, given as $T(m,n) = m^2 + mn + n^2$ (here m=3, n=1, T=13). In elongated icosahedra, the number Q serves to define the extent of elongation. Q, defined by coordinates M' and n', is any integer greater than T. The intermediate phage head is defined by (m'=3, n'=2, Q=17) and the normal head has (m'=3, n'=3, Q=21). (From Mosig, G. and Eiserling, F., *The Bacteriophages,* Vol. 2, Calendar, R., Ed., Plenum Press, New York, 1988, 521. With permission.)

MYOVIRIDAE

Models for DNA packing within the phage head

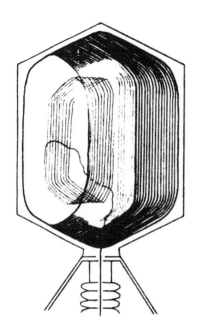

FIGURE 221
The axis of coiling is perpendicular to the phage axis. The outermost layer of DNA is likely to be laid down first; the last DNA to enter the capsid is located at the center of the coil and will be the first to be ejected. (From Earnshaw, W.C., King, J., Harrison, J., and Eiserling, F.A., *Cell*, 14, 559, 1978. © 1978 Cell Press, Cambridge, MA. With permission.)

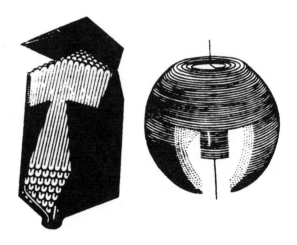

FIGURE 222
Alternatively, the DNA may be laid down in a "spiral-fold organization" of 180° folded rods or in a concentric shell with the center laid down first. (From Black, L.W., Newcomb, W.W., Boring, J.W., and Brown, J.C., *Proc. Natl. Acad. Sci. U.S.A.*, 82, 7960, 1985. With permission.)

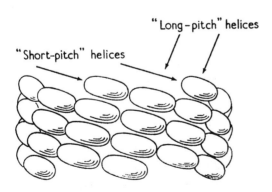

FIGURE 223
Arrangement of subunits in the contracted sheath of phage T4. Upon contraction, the subunits slide over each other, the tail sheath becomes shorter and thicker, and the number of cross striations decreases from 24 to 12. (From Moody, M.F., *J. Mol. Biol.*, 25, 167, 1967. With permission.)

FIGURE 224
Contracted tail sheath viewed from above, showing right-handed grooves forming a clockwise turning spiral. (From Moody, M.F., *J. Mol. Biol.*, 25, 167, 1967. With permission.)

MYOVIRIDAE

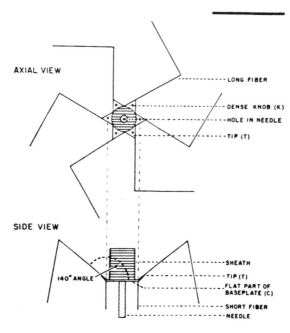

FIGURE 225
Base plate of T-even phage after sheath contraction; axial and side views. (From Simon, L.D. and Anderson, T.F., *Virology*, 32, 279, 1967. With permission.)

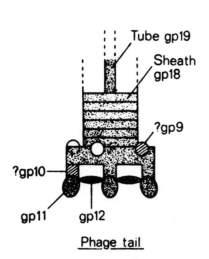

FIGURE 226
Tail and base plate proteins of T-even phages. The location of gp9 and gp10 was not established at the time of writing. (From Crowther, R.A., Lenk, E.V., Kikuchi, V., and King, J., *J. Mol. Biol.*, 116, 489, 1977. With permission.)

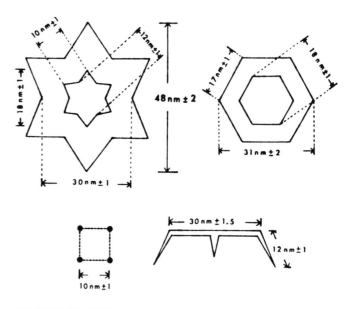

FIGURE 227
Dimensions of base plate structures of T-even phages. (From Cummings, D.J., Chapman, V.A., DeLong, S.S., Kusy, A.R., and Stone, K.R., *J. Virol.*, 6, 545, 1970. With permission.)

MYOVIRIDAE

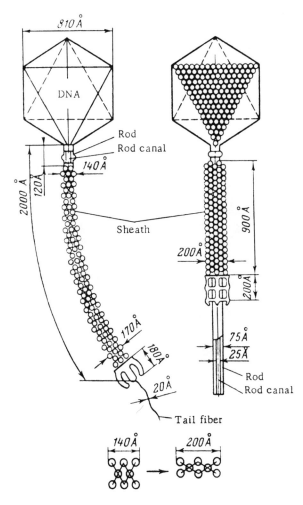

FIGURE 228
Phage No.1 of *Bacillus mycoides* with extended and contracted tail. (From Tikhonenko, A.S., *Ultrastructure of Bacterial Viruses*, Plenum Press, New York, 1970, 130. With permission.)

FIGURE 229
Phage SPO1 of *B. subtilis*. The head is shown with a T=16 lattice and is linked to the tail by a connector structure. The indistinct base plate has a six-fold symmetry. (From Eiserling, F.A., *Comprehensive Virology*, Fraenkel-Conrat, H. and Wagner, R.R., Eds., Plenum Press, New York, 1979, 543. With permission.)

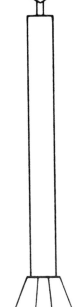

FIGURE 230
"Killer-particle" of *B. subtilis*, a defective temperate phage with a characteristically small head. Unable to multiply; kills bacteria from the outside. (Reprinted from Ackermann, H.-W. and Brochu, G., *CRC Handbook of Microbiology*, 2nd ed., Vol. 2, 1978, 691, CRC Press, Boca Raton, FL.)

MYOVIRIDAE

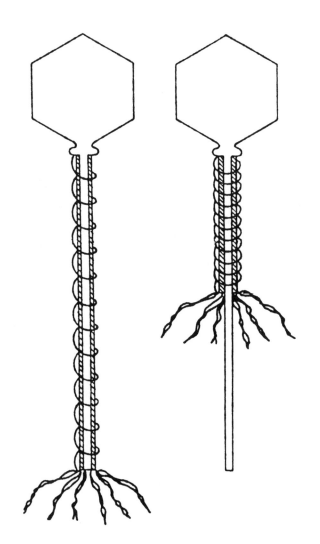

FIGURE 231

B. megaterium phage G, one of the largest phages known, characterized by a spiral filament around the tail; see also Figure 257. (From Ageno, M., Donelli, G., and Guglielmi, F., Structure and physico-chemical properties of bacteriophage G. II. The shape and symmetry of the capsid, *Micron*, 4, 376–403, © 1973 Elsevier Science Ltd., The Boulevard, Langford Lane, Kidlington OX5 1GB, UK. With permission.)

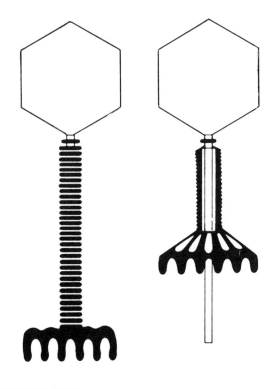

FIGURE 232

B. subtilis phage φ25 with extended and contracted tail; note the large base plate. (From Liljemark, W.F. and Anderson, D.L., *J. Virol.*, 6, 107, 1970. With permission.)

FIGURE 233

Gluconobacter phage A-1 (first described as an *Acetobacter* phage). (From Schocher, A.J., Kuhn, H., Schindler, B., Palleroni, N.J., Despreaux, C.W., Boublik, M., and Miller, P.A., *Arch. Microbiol.*, 121, 193, 1979. © Springer-Verlag. With permission.)

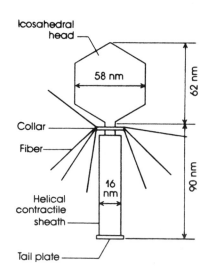

MYOVIRIDAE

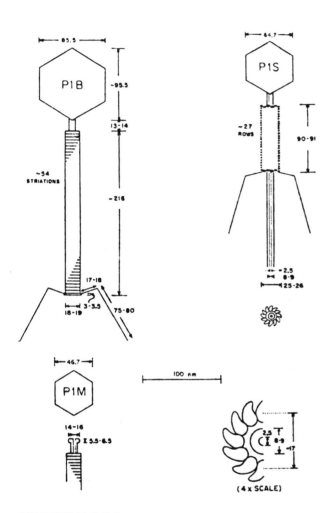

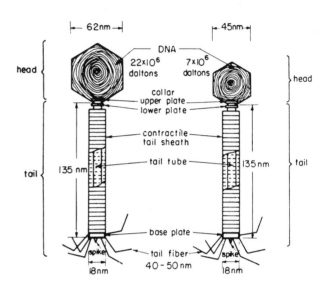

FIGURE 235
Coliphage P2 and its satellite P4. (From Goldstein, R., Lengyel, J., Pruss, G., Barrett, K., Calendar, R., and Six, E., *Curr. Topics Microbiol. Immunol.*, 68, 59, 1974. © Springer-Verlag. With permission.)

FIGURE 234
Coliphage P1, its morphological variants P1S and P1M, and structure of contracted tail. (From Walker, D.H. and Anderson, T.F., *J. Virol.*, 5, 765, 1970. With permission.)

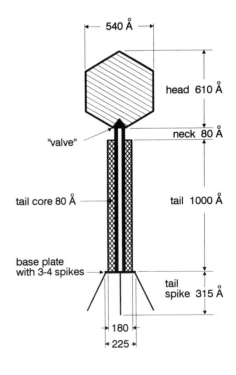

FIGURE 236
Coliphage Mu, here named Mu1. (By M. Côté, modified from Reference 213.)

**Tail long,
noncontractile**

II. SIPHOVIRIDAE

Heads are isometric or elongated. Tails are long and
thin (65 to 570 × 6 to 8 nm), noncontractile, and more
or less flexible. The family has about 2500 members and
is typified by coliphage λ.

Two early views of coliphage T1

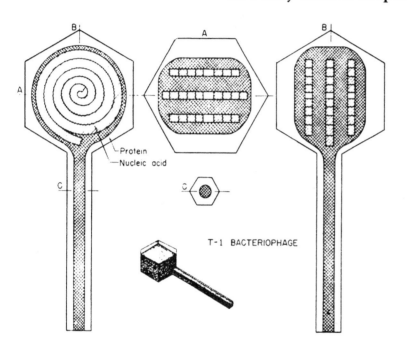

FIGURE 237
Model based on radiation experiments (1954). An outer protein
coat surrounds a threefold spiral of nucleic acid that is composed of
genetic material and nonessential "latent period factor". (From
Pollard, E., *Adv. Virus Res.,* 2, 109, 1954. With permission.)

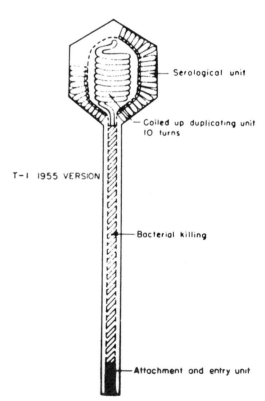

FIGURE 238
A slightly later model (1955), equally based
on radiation studies. (From Pollard, E.C.,
Hepatitis Frontiers, Hartman, F.W., LoGrippo,
G.A., Mateer, J.G., and Barron, J., Eds., p.
355. © 1957 Little, Brown and Company, Boston. With permission.)

SIPHOVIRIDAE

Three views of coliphage λ

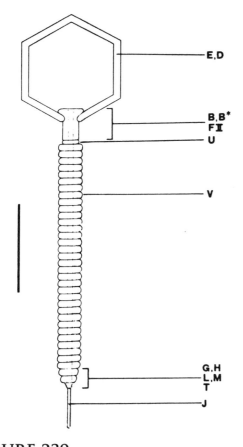

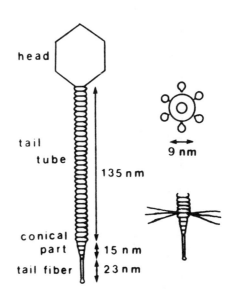

FIGURE 239

Location of phage components. The head contains two major capsid proteins, E and D. The head-tail connector comprises at least four proteins. The bar represents 50 nm. (From Eiserling, F.A., *Comprehensive Virology*, Vol. 13, Fraenkel-Conrat, H. and Wagner, R.R., Eds., Plenum Press, New York, 1979, 543. With permission.)

FIGURE 240

Complete phage, tail disk, and side fibers. (From Katsura, I., *Lambda II*, Hendrix, R.W., Roberts, J.W., Stahl, F.W., and Weisberg, R.A., Eds., Cold Spring Harbor Laboratory, Cold Spring Harbor, NY, 1983, 331. With permission.)

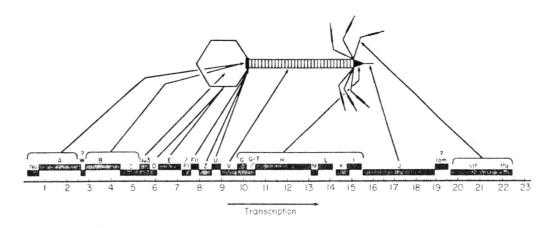

FIGURE 241

Size and location of morphopoietic genes and their site of action. (From Casjens, S., Hatfull, G., and Hendrix, R., *Semin. Virol.*, 3, 383, 1992. With permission.)

SIPHOVIRIDAE

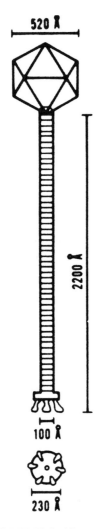

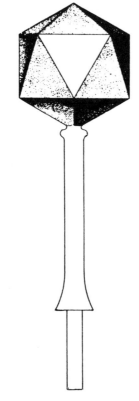

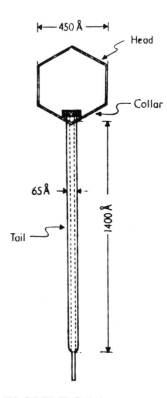

FIGURE 242

Bacillus subtilis phage φ105 and end-on view of base plate. (From Birdsell, D.C., Hathaway, G.M., and Rutberg, L., *J. Virol.*, 4, 264, 1969. With permission.)

FIGURE 243

B. megaterium phage α (1961). The "sheath" is a bundle of long collar fibers.[157] (Adapted from Chiozzotto, A., Coppo, A., Donini, P., and Graziosi, F., *Sci. Rep. Inst. Super. Sanita*, 1, 112, 1961. With permission.)

FIGURE 244

B. subtilis phage SPP1. (From Riva, S., Polsinelli, M., and Falaschi, A., *J. Mol. Biol.*, 35, 347, 1968. With permission.)

100 nm

FIGURE 245

Acinetobacter phage 531 showing tail disks and disk fibers. (By H.-W. Ackermann and M. Côté after Reference 221.)

SIPHOVIRIDAE

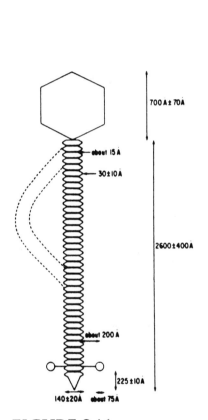

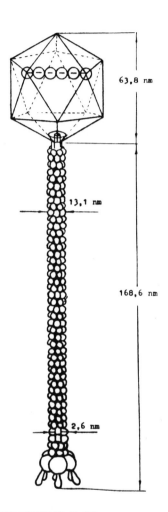

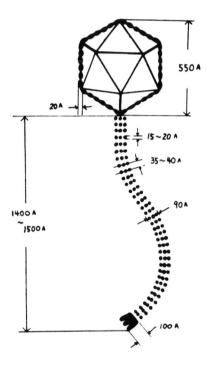

FIGURE 246
Agrobacterium phage PS-8. (From Knopf, U.C., *Arch. Ges. Virusforsch.*, 46, 205, 1974. With permission.)

FIGURE 247
Erwinia phage 59. (From Kishko, Ya. G., Ruban, V.I., Tovkach, F.I., Muraschchyk, I.G., and Danileychenko, V.V., *J. Virol.*, 46, 1018, 1983. With permission.)

FIGURE 248
Mycobacterium phage B-1. (From Takeya, K. and Amako, K., *Virology*, 24, 461, 1964. With permission.)

FIGURE 249
Lactobacillus phage LL-H. Bar represents 100 nm. (From Alatossova, T., *Acta Univ. Oulu*, Ser. A, No. 191, p. 34, 1987. With permission.)

**Tail short,
noncontractile**

III. PODOVIRIDAE

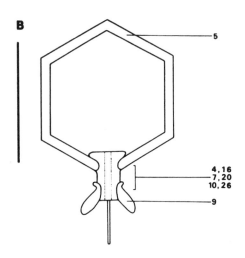

FIGURE 250

Salmonella phage P22 with location of structural proteins. (From Eiserling, F.A., *Comprehensive Virology,* Vol. 13, Fraenkel-Conrat, H. and Wagner, R.R., Eds., Plenum Press, New York, 1979, 543. With permission.)

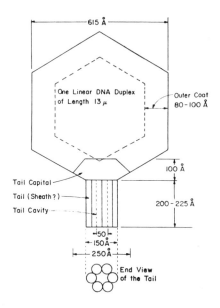

FIGURE 251

Cyanophage LPP-1. (From Luftig, R. and Haselkorn, R., *Virology,* 34, 664, 1968. With permission.)

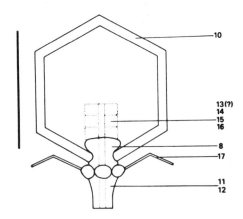

FIGURE 252

Coliphage T7. Proteins 13 to 16 are internal proteins. (From Eiserling, F.A., *Comprehensive Virology,* Vol. 13, Fraenkel-Conrat, H. and Wagner, R.R., Eds., Plenum Press, New York, 1979, 543. With permission.)

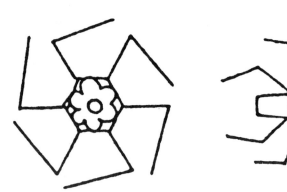

FIGURE 253

T7 tail and tail fibers. (From Matsuo-Kato, H., Fujisawa, H., and Minagawa, T., *Virology,* 109, 157, 1981. With permission.)

PODOVIRIDAE

B. subtilis phage φ29, a tiny phage with a complex structure

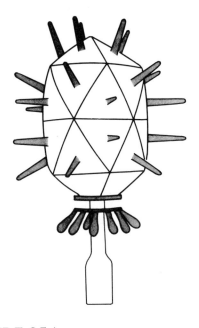

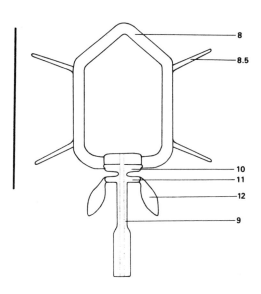

FIGURE 254
Outside view with possible location of head fibers. (By H.-W. Ackermann and M. Côté.)

FIGURE 255
Structure of phage and location of phage proteins. The upper collar is involved in head shape determination and the lower collar has a role in phage adsorption. The bar represents 50 nm. (From Eiserling, F.A., *Comprehensive Virology*, Vol. 13, Fraenkel-Conrat, H. and Wagner, R.R., Eds., Plenum Press, New York, 1979, 543. With permission.)

FIGURE 256
Head protein subunits are thought to be arranged in hexamers and pentamers. (From Hendrix, R.W., *Virus Structure and Assembly*, 169, Casjens, S., Ed., © 1985 Jones and Bartlett Publishers, Boston. Reprinted by permission.)

IV. MISCELLANY

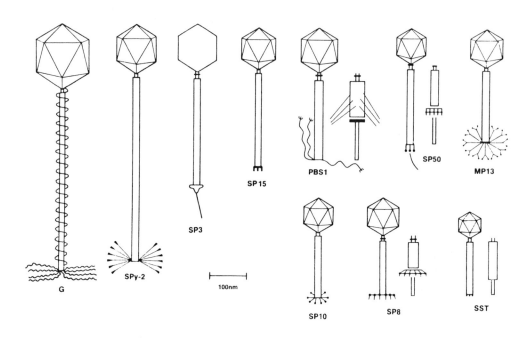

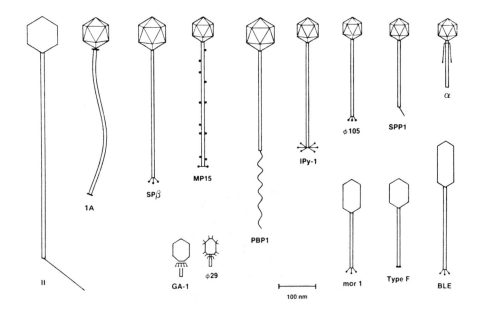

FIGURE 257

Bacillus phage species in 1987. Nine more species have been described since publication of this figure.[229] (From Ackermann, H.-W. and DuBow, M.S., *Viruses of Prokaryotes*, Vol. 2, 1987, 78 and 79, CRC Press, Boca Raton, FL.)

MISCELLANY

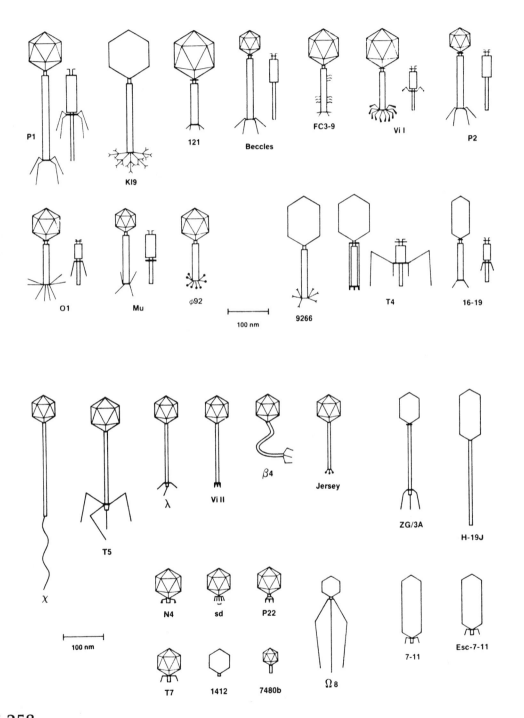

FIGURE 258
Enterobacterial phage species in 1987. As in bacilli, several new phage species have been found in the last years. (From Ackermann, H.-W. and DuBow, M.S., *Viruses of Prokaryotes*, Vol. 2, 1987, 96 and 97, CRC Press, Boca Raton, FL.)

MISCELLANY

Enterobacterial phages of different morphology

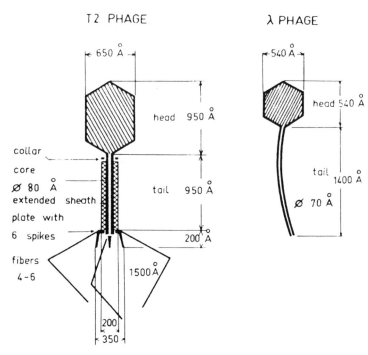

FIGURE 259
Dimensions of phages T2 and λ. (From Kellenberger, E., *Adv. Virus Res.*, 8, 1, 1961. With permission.)

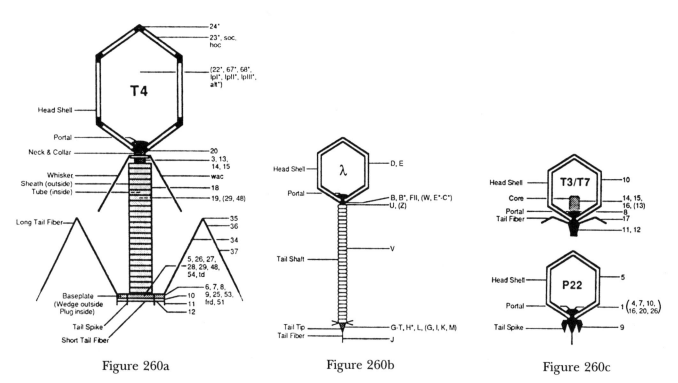

Figure 260a   Figure 260b   Figure 260c

FIGURE 260
Components and structural proteins of phages T4, λ, T3/T7, and P22. (From Casjens, S. and Hendrix, J., *The Bacteriophage*, Vol. 1, Calendar, R., Ed., Plenum Press, New York, 1987, 15. With permission.)

MISCELLANY

Types of *Salmonella* phages

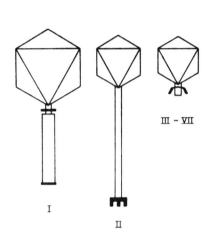

FIGURE 261

S. typhi phages Vi I to VII. (From Ackermann, H.-W., Berthiaume, L., and Kasatiya, S.S., *Can. J. Microbiol.,* 16, 411, 1970. With permission.)

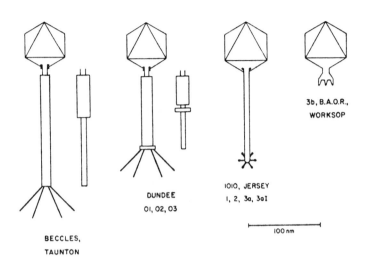

FIGURE 262

S. paratyphi B typing phages. Isometric phage heads were then widely thought to be octahedra. (From Ackermann, H.-W., Berthiaume, L., and Kasatiya, S.S., *Can. J. Microbiol.,* 18, 77, 1972. With permission.)

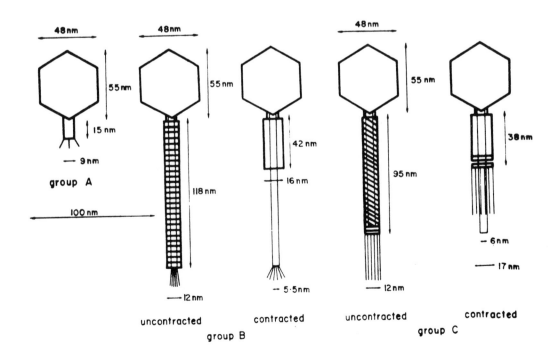

FIGURE 263

Three types of *S. potsdam* phages. Tail fiber representations do not indicate their actual length and numbers. (From Nutter, R.L., Bullas, L.R., and Schultz, R.L., *J. Virol.,* 5, 754, 1970. With permission.)

MISCELLANY

Contractile defective phages

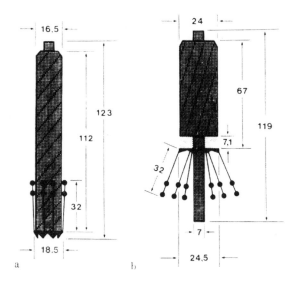

FIGURE 264

A tail-like bacteriocin of *Rhizobium* in (a) extended and (b) contracted state. (From Lotz, W. and Mayer, F., *J. Virol.*, 9, 160, 1972. With permission.)

FIGURE 265

Two similar types of particles from *Chromobacterium;* biological activity unknown. (From Ackermann, H.-W. and Gauvreau, L., *Zentralbl. Bakteriol. Parasitenk. Infektionskr. Hyg. Abt. I Orig. A,* 221, 196, 1972. With permission.)

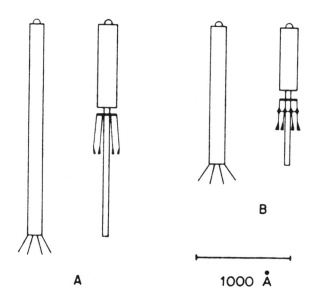

A

B

1000 Å

MISCELLANY

Models of tail structure

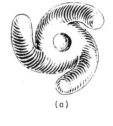

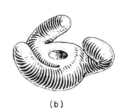

(a)

(b)

(c)

(d)

FIGURE 266

Tail structure of *Caulobacter* phage φCbK. a to c. Views of one enantiomer for a threefold annulus. d. Helix of 16 annuli stacked at appropriate angles. The projecting arms of the annuli are parallel to each other and produce steep helical grooves. The phage head would be attached at the top of the helix. (From Leonard, K.R., Kleinschmidt, A.K., and Lake, J.A., *J. Mol. Biol.*, 81, 249, 1973. With permission.)

	lambda tail	T4 tube
outer (inner diameter)	17 (9) nm	9 nm
central channel	3 - 3.5 nm	3.2 -3.8 nm
length (no. of disks)	135 nm (32 x 4.2 nm)	98.4 nm (24 x 4.1 nm)
template molecule length (Mr)	gp H 853 aa ($93,393 \rightarrow 79,000$)	gp 29 591 aa (64,382)
number	3-7/tail	4-7/tail
tube/template	0.158 nm/aa	0.167 nm/aa

FIGURE 267

Comparison of the tail of phage λ and the tail tube of T4, showing similar properties of potential tail length-determining "ruler" proteins. (From Eiserling, F.A., *Bacteriophage T4*, Mathews, C.K., Kutter, E.M., Mosig, G., and Berget, P.B., Eds., American Society for Microbiology, Washington, DC, 1983, 11. With permission.)

MISCELLANY

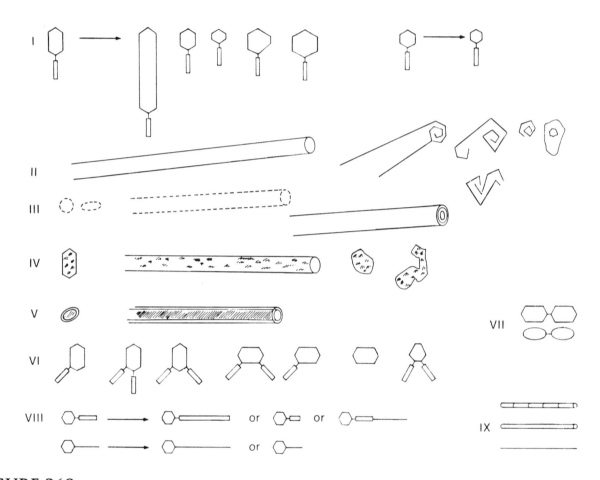

FIGURE 268

Morphological aberrations in tailed phages. I. Derivatives of normal heads. II. Smooth polyheads. III. Rough heads and polyheads. IV. Mottled heads and polyheads. V. t particles and possible derivatives. VI. Abnormal tail insertion. VII. Double heads. VIII. Elongated and shortened tails with or without sheath. IX. Polysheaths and polytubes. (From Ackermann, H.-W. and DuBow, M.S., *Viruses of Prokaryotes*, Vol. 2, 1987, 73, CRC Press, Boca Raton, FL.)

Tailed phages produce a wide variety of morphological aberrations, depending on the complexity of the phage, genetic defects, timing of phage assembly, and composition of growth media. For example, T-even phages produce several variants of the normal capsid, e.g., extremely long "giant" heads and or shortened, isometric, biprolate, or tridimensionally enlarged heads. All of them may be filled with DNA. Some abnormal structures, e.g., rough heads, represent proheads that have not been converted to normal capsids. Aberrant particles often provide clues for the identification of phage genes.

MISCELLANY

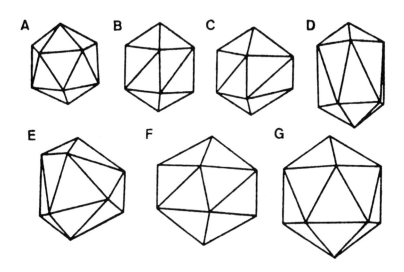

FIGURE 269
Aberrant heads in phage T4. D represents a normal capsid. (From Keller, B., Dubochet, J., Adrian, M., Maeder, M., Wurtz, M., and Kellenberger, E., *J. Virol.*, 62, 2960, 1988. With permission.)

FIGURE 270
T4 head polymorphism. Multi- and single-layered tubes and τ-particles are made with unmodified protein 23. None of these structures contains DNA. Isometric ("petite"), normal (prolate), and giant heads are made with cleaved protein 23^c. They contain normally DNA amounts directly proportional to their volume. (From Kellenberger, E., *Generation of Subcellular Structure*, Markham, R., Ed., Elsevier, Amsterdam, 1973, 62. With permission.)

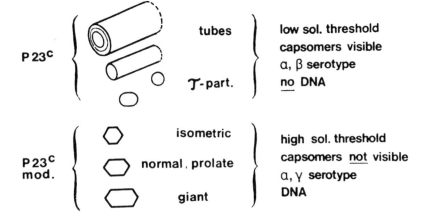

GLOSSARY AND ABBREVIATIONS

Arbovirus A virus replicating in both an arthropod and a vertebrate ("arthropod-borne").

Bacteriophage A virus replicating in bacteria only; also called a "phage."

Binary symmetry Combination of cubic and helical symmetry.

Capsid Protein shell which surrounds nucleic acid, nucleoprotein, or core.

Capsomer(e) Morphological unit forming the capsid, visible in the electron microscope; composed of structural units (protomers).

Core Internal body enclosed in a capsid or envelope; is or contains a nucleoprotein complex. Also the inner tube of contractile phage tails.

Cubic symmetry Form of capsid in which protein subunits are assembled to a compact shell with three axes of symmetry, passing through apices, edges, and faces.

Dodecahedron One of the five Platonic solids, characterized by 12 pentagonal faces.

ds Double-stranded.

Envelope Lipoprotein membrane derived from a host cell membrane or synthetized *de novo*, surrounds the capsid or nucleocapsid, usually a bilayer carrying virus-specified proteins or peplomers.

gp Gene product.

Helical symmetry Form of capsid in which protein subunits are assembled to a spiral; the only axis of symmetry is the length axis of the particle.

Hexon Capsomer with six subunits, surrounded by six identical capsomers, said to be "hexavalent."

Icosahedron One of the five Platonic solids; has 20 identical triangular faces, 30 edges, 12 apices, and two-, three-, and five-fold axes of symmetry.

Inclusion body
1. Virus-coded intracellular body containing protein and viruses.
2. Any array of viruses or assembly of abnormal protein detectable by light microscopy.

K Kilodalton (kDa).

-mer Suffix derived from Greek *meros*, part; e.g., dimer, trimer, pentamer.

Naked Colloquial for "nonenveloped."

Nucleocapsid Nucleic acid included by protein.

Nucleoid Electron-dense central region of some viruses.

Nucleosome Bead-like arrangement of nucleocapsid.

-partite   Suffix meaning "consisting of parts"; e.g., bipartite, tripartite.

Occluded   A virus producing an occlusion body.

Occlusion body   Large virus-coded protein crystal containing viruses, preferred to the more general term "inclusion body."

Penton   Capsomer with five subunits, usually surrounded by five hexons, said to be "pentavalent."

Peplomer   Glycoprotein projection (spike, knob) located on the envelope.

Polyhedron   Large, many-sided or rounded occlusion body.

Skewed   Arrangement of rows of capsomers in a left-handed (laevo, *l*) or right-handed (dextro, *d*) turn.

ss   Single-stranded.

Structural unit   Protein subunit or "protomer," the basic building block of the capsid.

Symmetry   Structural principle of viral capsid or nucleocapsid.

Triangulation number Number of identical equilateral triangles composing a face of a given icosahedron, e.g., T=4, T=9, T=16.

Virion   Complete infectious virus particle.

Viroid   Rod-shaped infectious ssRNA of low molecular weight.

VP   Viral protein.

+ sense   Plus-sense or positive-sense. In RNA viruses, the strand which functions as a messenger (mRNA).

– sense   Minus-sense or negative-sense: nucleic acid complementary to the + strand.

REFERENCES

1. **Doane, F.W. and Anderson, N.,** *Electron Microscopy in Diagnostic Virology. A Practical Guide and Atlas*, Cambridge University Press, Cambridge, UK, 1987, 48.
2. **Francki, R.I.B., Milne, R.G., and Hatta, T.,** *Atlas of Plant Viruses*, Vol. 1, CRC Press, Boca Raton, FL, 1985, 9.
3. **Kurstak, E., Tijssen, P., and Garzon, S.,** Densonucleosis viruses (Parvoviridae), in *The Atlas of Insect and Plant Viruses*, Maramorosch, K., Ed., Academic Press, New York, 1977, 67.
4. **Palmer, E.L. and Martin, M.L.,** *Electron Microscopy in Viral Diagnosis*, CRC Press, Boca Raton, FL, 1988, 2.
5. **Van Kammen, A. and Mellema, J.E.,** Comoviruses, in *The Atlas of Plant and Insect Viruses*, Maramorosch, K., Ed., Academic Press, New York, 1977, 167.
6. **Casjens, S.,** An introduction to virus structure and assembly, in *Virus Structure and Assembly*, Casjens, S., Ed., Jones and Bartlett, Boston, 1985, 1.
7. **Caspar, D.L.D. and Klug, A.,** Physical principles in the construction of regular viruses, *Cold Spring Harbor Symp. Quant. Biol.*, 27, 1, 1962.
8. **Harrison, S.C.,** Principles of virus structure, in *Virology*, 2nd ed., Vol. 1, Fields, B.N. and Knipe, D.M., Eds., Raven Press, New York, 1990, 37.
9. **Horne, R.W.,** *Virus Structure*, Academic Press, New York, 1974, 7, 17, and 28.
10. **Nermut, M.V.,** General principles of virus architecture, in *Animal Virus Structure*, Nermut, M.V. and Steven, A.C., Eds., Elsevier, Amsterdam, 1987, 3.
11. **Rayment, I.,** Animal virus structure, in *Biological Macromolecules and Assemblies*, Vol. 1, Jurnak, F.A. and McPherson, A., Eds., John Wiley & Sons, New York, 1984, 255.
12. **Murphy, F.A., Fauquet, C.M., Bishop, D.H.L., Ghabrial, S.A., Jarvis, A.W., Martelli, G.P., Mayo, M.A., and Summers, M.D.,** Eds., *Virus Taxonomy, The Classification and Nomenclature of Viruses, Sixth Report of the International Committee on Taxonomy of Viruses, Arch. Virol.* (Suppl. 3), Springer-Verlag Vienna, in press.
13. **Matthews, R.E.F.,** A history of viral taxonomy, in *A Critical Appraisal of Viral Taxonomy*, Matthews, R.E.F., Ed., CRC Press, Boca Raton, FL, 1983, 1.
14. **Lwoff, A., Horne, R.W., and Tournier, P.,** A system of viruses, *Cold Spring Harbor Symp. Quant. Biol.*, 27, 51, 1962.
15. **Van Regenmortel, M.H.V.,** Virus species, a much overlooked but essential concept in virus classification, *Intervirology*, 31, 241, 1990.
16. **Fenner, F.J., Gibbs, E.P.J., Murphy, F.A., Rott, R., Studdert, M.J., and White, D.O.,** *Veterinary Virology*, 2nd ed., Academic Press, San Diego, 1993, 21.
17. **Bellett, A.J.D., Fenner, F., and Gibbs, A.J.,** The viruses, in *Frontiers of Biology*, Vol. 31, Gibbs, A.J., Ed., North-Holland Publishing Co., Amsterdam, 1973, 41.
18. **Francki, R.I.B., Fauquet, C.M., Knudson, D.L., and Brown, F.,** Eds., *Classification and Nomenclature of Viruses, Fifth Report of the International Committee on Taxonomy of Viruses, Arch. Virol.* (Suppl. 2), Springer-Verlag, Vienna, 1991, 60.
19. **Bradley, D.E.,** The morphology and physiology of bacteriophages as revealed in the electron microscope, *J. R. Microscop. Soc.*, 84, 275, 1965.
20. **Ackermann, H.-W. and DuBow, M.S.,** *Viruses of Prokaryotes*, Vol. 1, 1987, 16 and 73.
21. **Bernhard, W.,** The detection and study of tumor viruses with the electron microscope, *Cancer Res.*, 20, 712, 1960.
22. **Horne, R.W., Pasquali Ronchetti, I., and Hobart, J.M.,** A negative staining-carbon film technique for studying viruses in the electron microscope, II. Applications to adenovirus type 5, *J. Ultrastruct. Res.*, 51, 233, 1975.
23. **Russell, W.C., Hayashi, K., Sanderson, P.J., and Pereira, H.G.,** Adenovirus antigens — a study of their properties and sequential development in infection, *J. Gen. Virol.*, 1, 495, 1967.
24. **Everitt, E., Lutter, L., and Philipson, L.,** Structural proteins of adenoviruses. XII. Location and neighbor relationship among proteins of adenovirion type 2 as revealed by enzymatic iodination, immunoprecipitation and chemical cross-linking, *Virology*, 67, 197, 1975.
25. **Ginsberg, H.S.,** Adenovirus structural proteins, in *Comprehensive Virology*, Vol. 13, Fraenkel-Conrat, H. and Wagner, R.R., Eds., Plenum Press, New York, 1979, 49.

26. **Nermut, M.V.,** The architecture of adenoviruses: recent views and problems, *Arch. Virol.,* 64, 175, 1980.

27. **Nermut, M.V.,** Adenoviridae, in *Animal Virus Structure,* Nermut, M.V. and Steven, A.C., Eds., Elsevier, Amsterdam, 1987, 373.

28. **Nermut, M.V.,** Structural elements in adenovirus cores. Studies by means of freeze-fracturing and ultrathin sectioning, *Arch. Virol.,* 57, 323, 1978.

29. **Ginsberg, H.S. and Young, C.S.H.,** Genetics of adenoviruses, in *Comprehensive Virology,* Vol. 9, Fraenkel-Conrat, H. and Wagner, R.R., Eds., Plenum Press, New York, 1977, 27.

30. **Berger, J., Burnett, R.M., and Grütter, M.,** Small angle X-ray scattering studies on adenovirus type 2 hexon, *Biochim. Biophys. Acta,* 535, 233, 1978.

31. **Young, P.R.,** Arenaviridae, in *Animal Virus Structure,* Nermut, M.V. and Steven, A.C., Eds., Elsevier, Amsterdam, 1987, 185.

32. **Bishop, D.H.L.,** Lymphocystic choriomeningitis virus ambisense coding: a strategy for persistent infections? in *Immunobiology and Pathogenesis of Persistent Virus Infections,* Lopez, C., Ed., American Society for Microbiology, Washington, DC, 1988, 79.

33. **Horzinek, C.,** *Kompendium der allgemeinen Virologie,* 2nd ed., Paul Parey, Berlin, 1985, 5, 47, 49, 51, 58, 67, and 72.

34. **Rohrmann, G.F.,** Baculovirus structural proteins, *J. Gen. Virol.,* 73, 749, 1992.

35. **David, W.A.L.,** The granulosis virus of *Pieris brassicae* (L.) and its relationship with its host, *Adv. Virus Res.,* 22, 111, 1978.

36. **Kawamoto, F. and Asayama, T.,** Studies on the arrangement patterns of nucleocapsids within the envelopes of nuclear-polyhedrosis viruses in the fat-body cells of the brown tail moth, *Euproctis similis, J. Invert. Pathol.,* 26, 47, 1975.

37. **Bergold, G.H.,** Fine structure of some insect viruses, *J. Insect Pathol.,* 5, 11, 1963.

38. **Hughes, K.M.,** Fine structure and development of two polyhedrosis viruses, *J. Invert. Pathol.,* 19, 198, 1972.

39. **Federici, B.A. and Humber, R.A.,** A possible baculovirus in the insect-parasitic fungus, *Strongwellsea magna, J. Gen. Virol.,* 35, 387, 1977.

40. **Harrap, K.A.,** The structure of nuclear polyhedrosis viruses. I. The inclusion body, *Virology,* 50, 114, 1972.

41. **Pettersson, R.F. and von Bonsdorff, C.-H.,** Bunyaviridae, in *Animal Virus Structure,* Nermut, M.V. and Steven A.C., Eds., Elsevier, Amsterdam, 1987, 147.

42. **Bishop, D.H.L.,** Genetic potential of bunyaviruses, *Curr. Topics Microbiol. Immunol.,* 86, 1, 1979.

43. **Goldbach, R. and De Haan, P.,** Prospects of engineered forms of resistance against tomato spotted wilt virus, *Semin. Virol.,* 4, 381, 1993.

44. **Best, J.R.,** Tomato spotted wilt virus, *Adv. Virus Res.,* 13, 65, 1968.

45. **Garwes, D.J.,** Structure and physicochemical properties of coronaviruses, in *Entérites virales chez l'homme et l'animal,* Colloques Vol. 90, Bricout, F. and Scherrer, R., Eds., Editions INSERM, Paris, 1979, 141.

46. **Macnaughton, M.R. and Davies, H.A.,** Coronaviridae, in *Animal Virus Structure,* Nermut, M.V. and Steven, A.C., Eds., Elsevier, Amsterdam, 1987, 173.

47. **Holmes, K.V.,** Coronaviridae and their replication, in *Virology,* 2nd ed., Vol. 1, Fields, B.N. and Knipe, D.M., Eds., Raven Press, New York, 1990, 841.

48. **Weiss, M. and Horzinek, M.C.,** The proposed family *Toroviridae:* agents of enteric infections, *Arch. Virol.,* 92, 1, 1987.

49. **Horzinek, M.C., Ederveen, J., Kaeffer, B., De Boer, D., and Weiss, M.,** The peplomers of Berne virus, *J. Gen. Virol.,* 67, 2475, 1986.

50. **Snijder, E.J. and Horzinek, M.C.,** Toroviruses: replication, evolution and comparison with other members of the coronavirus-like superfamily, *J. Gen. Virol.,* 74, 2305, 1993.

51. **Cavanagh, D., Brien, D.A., Brinton, M., Enjuanes, L., Holmes, K.V., Horzinek, M.C., Lai, M.M.C., Laude, H., Plagemann, P.G.W., Siddell, S., Spaan, W.J.M., Taguchi, F., and Talbot, P.J.,** Revision of the taxonomy of the *Coronavirus, Torovirus* and *Arterivirus* genera, *Arch. Virol.,* 135, 227, 1994.

52. **Schäfer, R., Hinnen, R., and Franklin, R.M.,** Structure and synthesis of a lipid-containing bacteriophage. Properties of the structural proteins and distribution of the phospholipid, *Eur. J. Biochem.,* 50, 15, 1974.

53. **Gonzalez, C.F., Langenberg, W.G., Van Etten, J.L., and Vidaver, A.K.,** Ultrastructure of bacteriophage ø6: arrangement of the double-stranded RNA and envelope, *J. Gen. Virol.,* 35, 353, 1977.

54. **Mindich, L.,** Bacteriophage ø6: a unique virus having a lipid-containing membrane and a genome composed of three dsRNA segments, *Adv. Virus Res.,* 35, 137, 1988.

55. **Heinz, F.X.,** Epitope mapping of flavivirus glycoproteins, *Adv. Virus Res.,* 31, 103, 1986.

56. **Stevens, W.A.,** *Virology of Flowering Plants,* Blackie & Sons, Glasgow, 1983, 79, 80, and 89.

57. **Stannard, L.M.,** Hepadnaviridae, in *Animal Virus Structure,* Nermut, M.V. and Steven, A.C., Eds., Elsevier, Amsterdam, 1987, 361.

58. **Ackermann, H.-W., Cherchel, G., Valet, J.-P., Matte, J., Moorjani, S., and Higgins, R.,** Expériences sur la nature de particules trouvées dans des cas d'hépatite virale: type coronavirus, antigène Australia et particules de Dane, *Can. J. Microbiol.,* 20, 193, 1974.

59. **Neurath, A.R., Jameson, B.A., and Huima, T.,** Hepatitis B virus proteins eliciting protective immunity, *Microbiol. Sci.,* 4, 48, 1987.

60. **Lutwick, L.I.,** Hepatitis B virus, in *Textbook of Human Virology,* 2nd ed., Belshe, R.B., Ed., Mosby-Year Book, St. Louis, 719, 1991.

61. **Blum, H.E.,** Hepatitis B virus: significance of naturally occurring mutants, *Intervirology,* 35, 40, 1993.

62. **Longson, M.,** Herpes simplex, in *Principles and Practice of Clinical Virology,* 2nd ed., Zuckerman, A.J., Banatvala, J.E., and Pattison, J.E., Eds., John Wiley & Sons, Chichester, U.K., 1990, 3.

63. **Hay, J. Roberts, C.R., Ruyechan, W.T., and Steven, A.C.,** Herpesviridae, in *Animal Virus Structure,* Nermut, M.V. and Steven, A.C., Eds., Elsevier, Amsterdam, 1987, 391.

64. **Huraux, J.M., Nicolas, J.C., and Agut, H.,** *Virologie,* Flammarion Médecine-Sciences, Paris, 1985, 69.

65. **McKendrick, G.D.W. and Sutherland, S.,** *An Introduction to Herpes Virus Infections,* Wellcome Foundation, London, 1983, 10.

66. **Horne, R.W.,** *The Structure and Function of Viruses,* Edward Arnold, London, 1978, 7 and 49.

67. **Roizman, B. and Furlong, D.,** The replication of herpesviruses, in *Comprehensive Virology,* Vol. 3, Fraenkel-Conrat, H. and Wagner, R.R., Eds., Plenum Press, New York, 1974, 229.

68. **Nazerian, K.,** DNA configuration in the core of Marek's disease virus, *J. Virol.,* 13, 1148, 1974.

69. **Webster, R.E., Grant, R.A., and Hamilton, L.A.W.,** Orientation of the DNA in the filamentous bacteriophage fl, *J. Mol. Biol.,* 152, 357, 1981.

70. **Bruce, J., Gourlay, R.N., Hull, R., and Garwes, D.J.,** Ultrastructure of Mycoplasmatales virus laidlawii I, *J. Gen. Virol.,* 16, 215, 1972.

71. **Webster, R.E. and Lopez, J.,** Structure and assembly of the class I filamentous bacteriophage, in *Virus Structure and Assembly,* Casjens, S., Ed., Jones and Bartlett, Boston, 1985, 235.

72. **Makowski, L., Caspar, D.L.D., and Marvin, D.A.,** Filamentous bacteriophage PF1 structure determined at 7 Å resolution by refinement of models for the a-helical subunit, *J. Mol. Biol.,* 140, 149, 1980.

73. **Wrigley, N.G.,** An electron microscope study of the structure of *Sericesthis iridescent* virus, *J. Gen. Virol.,* 5, 123, 1969.

74. **Madeley, C.R., Smail, D.A., and Egglestone, S.I.,** Observations on the fine structure of lymphocystis virus from European flounders and plaice, *J. Gen. Virol.,* 40, 421, 1978.

75. **Berthiaume, L., Alain, R., and Robin, J.,** Morphology and ultrastructure of Lymphocystis disease virus, a fish iridovirus, grown in tissue culture, *Virology,* 135, 10, 1984.

76. **Heppell, J. and Berthiaume, L.,** Ultrastructure of lymphocystis disease virus (LDV) as compared to frog virus 3 (FV$_3$) and chilo iridescent virus (CIV): effects of enzymatic digestions and detergent degradations, *Arch. Virol.,* 125, 215, 1992.

77. **Darcy, F. and Devauchelle, G.,** Iridoviridae, in *Animal Virus Structure,* Nermut, M.V. and Steven, A.C., Eds., Elsevier, Amsterdam, 1987, 407.

78. **Darcy-Tripier, F., Nermut, M.V., Braunwald, J., and Williams, L.D.,** The organisation of frog virus 3 as revealed by freeze-etching, *Virology,* 138, 287, 1984.

79. **Goodheart, C.R.,** *An Introduction to Virology,* W.B. Saunders, Philadelphia, 1969, 37.

80. **Paranchych, W.,** Attachment, ejection and penetration stages of the RNA phage infectious process, in *RNA Phages,* Zinder, N.D., Ed., Cold Spring Harbor Laboratory, Cold Spring Harbor, NY, 1975, 85.

81. **Valegård, K., Liljas, L., Fridborg, K., and Unge, T.,** The three-dimensional structure of the bacterial virus MS2, *Nature,* 344, 36, 1990.

82. **Zillig, W.R., Reiter, H.-D., Palm, P., Gropp, P., Neumann, H., and Rettenberger, M.,** Viruses of archaebacteria, in *The Bacteriophages,* Vol. 1, Calendar, R., Ed., Plenum Press, New York, 1987, 517.

83. **Zillig, W.,** Viruses of archaebacteria, in *Virus Taxonomy. The Classification and Nomenclature of Viruses, Sixth Report of the International Committee on Taxonomy of Viruses,* Murphy, F.A., Fauquet, C.M., Bishop, D.H.L., Ghabrial, S.A., Jarvis, A.W., Martelli, G.P., Mayo, M.A., and Summers, M.D., Eds., *Arch Virol.* (Suppl. 3), Springer-Verlag Vienna, in print.

84. **Hayashi, M.,** Morphogenesis of the isometric phages, in *The Single-Stranded DNA Phages,* Denhardt, D.T., Dressler, D., and Ray, D.S., Eds., Cold Spring Harbor Laboratory, Cold Spring Harbor, NY, 531, 1978.

85. **Rott, R.,** Molecular basis of infectivity and pathogenicity of myxovirus, *Arch. Virol.*, 59, 285, 1979.

86. **Scheid, H.,** Paramyxoviridae, in *Animal Virus Structure*, Nermut, M.V. and Steven, A.C., Eds., Elsevier, Amsterdam, 1987, 233.

87. **Lund, G.A., Tyrrell, D.L.J., Bradley, R.D., and Scraba, D.G.,** The molecular length of measles virus nucleocapsids, *J. Gen. Virol.*, 65, 1535, 1984.

88. **Kingsbury, D.W.,** Paramyxoviridae and their replication, in *Virology*, 2nd ed., Vol. 1, Fields, B.N. and Knipe, D.M., Eds., Raven Press, New York, 1990, 945.

89. **Vainionpää, R., Marusyk, R., and Salmi, A.,** The Paramyxoviridae: aspects of molecular structure, pathogenesis, and immunity, *Adv. Virus Res.*, 37, 211, 1989.

90. **Brown, J.C. and Newcomb, W.W.,** Rhabdoviridae, in *Animal Virus Structure*, Nermut, M.V. and Steven, A.C., Eds., Elsevier, Amsterdam, 1987, 199.

91. **Bradish, C.J. and Kirkham, J.B.,** The morphology of vesicular stomatitis virus (Indiana C) derived from chick embryos or cultures of BHK21/13 cells, *J. Gen. Microbiol.*, 44, 359, 1966.

92. **Wunner, W.H., Larson, J.K., Dietzschold, B., and Smith, C.L.,** The molecular biology of rabies virus, *Rev. Infect. Dis.*, 10(Suppl. 4), 771, 1988.

93. **Simpson, R.W. and Hauser, R.E.,** Structural components of vesicular stomatitis virus, *Virology*, 29, 654, 1966.

94. **Vernon, S.K., Neurath, A.R., and Rubin, B.A.,** Electron microscopy studies on the structure of rabies virus, *J. Ultrastruct. Res.*, 41, 29, 1972.

95. **Francki, R.I.B. and Randles, J.W.,** Rhabdoviruses infecting plants, in *Rhabdoviruses*, Vol. 3, Bishop, D.H.L., Ed., CRC Press, Boca Raton, FL, 135, 1980.

96. **Francki, R.I.B.,** Plant rhabdoviruses, *Adv. Virus Res.*, 18, 257, 1973.

97. **Howatson, A.F.,** Vesicular stomatitis and related viruses, *Adv. Virus Res.*, 16, 195, 1970.

98. **Mammette, A.,** *Virologie médicale à l'usage des étudiants en médecine*, 14th ed., Editions C. et R., La Madeleine, France, 1992, 350.

99. **Kaplan, M.M. and Webster, R.G.,** The epidemiology of influenza, *Sci. Am.*, 237(6), 88, 1977.

100. **Blaskovic, D.,** Virus infection of the cell, *Acta Virol.*, 3(Suppl.), 7, 1959.

101. **Kates, M., Allison, A.C., Tyrell, D.A.J., and James, A.T.,** Origin of lipids in influenza virus, *Cold Spring Harbor Symp. Quant. Biol.*, 27, 193, 1962.

102. **Oxford, J.S. and Hockley, D.J.,** Orthomyxoviridae, in *Animal Virus Structure*, Nermut, M.V. and Steven, A.C., Eds., Elsevier, Amsterdam, 213, 1987.

103. **Apostolov, K. and Flewett, T.H.,** Further observations on the structure of influenza viruses A and C, *J. Gen. Virol.*, 4, 365, 1969.

104. **Aymard, M.,** Les orthomyxoviridés: les virus grippaux, in *Virologie médicale*, Maurin, J., Ed., Flammarion Médecine-Sciences, Paris, 1985, 448.

105. **Schulze, I.T.,** Structure of the influenza virion, *Adv. Virus Res.*, 18, 1, 1973.

106. **Palese, P. and Ritchey, M.D.,** Myxoviridae: orthomyxoviruses–influenzaviruses, in *Virology and Rickettsiology*, Vol. 1, Part I, Hsiung, G.-D. and Green, R.H., Eds., CRC Press, Boca Raton, FL, 1978, 337.

107. **Compans, R.W., Content, J., and Duesberg, P.H.,** Structure of the ribonucleoprotein of influenza virus, *J. Virol.*, 10, 795, 1972.

108. **Horne, R.W.,** The structure of viruses, *Sci. Am.*, 108 (1), 48, 1963.

109. **Hogle, J.M., Chow, M., and Filman, D.J.,** The structure of poliovirus, *Sci. Am.*, 256 (3), 42, 1987.

110. **Rueckert, R.R.,** Picornaviridae and their replication, in *Virology*, 2nd ed., Vol. 1, Fields, B.N. and Knipe, D.M., Eds., Raven Press, New York, 1990, 507.

111. **Chen, Zh., Stauffacher, C., Schmidt, T., Fisher, A., and Johnson, J.E.,** RNA packaging in bean pod mottle virus, in *New Aspects of Positive-Strand RNA Viruses*, Brinton, M.A. and Heinz, F.X., Eds., American Society for Microbiology, Washington, DC, 1990, 218.

112. **Müller, G. and Williamson, J.D.,** Poxviridae, in *Animal Virus Structure*, Nermut, M.V. and Steven, A.C., Eds., Elsevier, Amsterdam, 1987, 421.

113. **Peters, D.,** Morphology of resting vaccinia virus, *Nature*, 178, 1453, 1956.

114. **Peters, D.,** Struktur und Entwicklung der Pockenviren, in *Proc. Fourth Int. Congr. Electron Microsc.*, Berlin, Sept. 10-17, Vol. 2, Bargmann, W., Peters, D., and Wolpers, C., Eds., Springer-Verlag, Berlin, 1958, 552.

115. **Friend Norton, C.,** *Microbiology*, Addison-Wesley, Reading, MA, 1981, 306.

116. **Avakyan, A.A. and Byrovsky, A.F.,** The structure of intracellular variola virus, *Acta Virol.*, 8, 481, 1964.

117. **Westwood, J.C.N., Harris, W.J., Zwartouw, H.T., Titmuss, D.H.J., and Appleyard, G.,** Studies on the structure of vaccinia virus, *J. Gen. Microbiol.*, 34, 67, 1964.

118. **Vilaginès, P. and Vilaginès, R.,** Les poxviridés: caractères généraux, in *Virologie médicale*, Maurin, J., Ed. Flammarion Médecine-Sciences, Paris, 1985, 310.

119. **Byrovsky, A.F.,** Changes in the envelope of vaccinia virus during ontogenesis, *Acta Virol.*, 8, 490, 1964.

120. **Nagington, J., Newton, A.A., and Horne, R.W.,** The structure of Orf virus, *Virology*, 23, 461, 1964.

121. **James, M.H. and Peters, D.,** The organization of nucleoprotein within fowlpox virus, *J. Ultrastruct. Res.*, 35, 626, 1971.

122. **Bergoin, M., Devauchelle, G., and Vago, C.,** Electron microscopic study of *Melolontha* poxvirus: the fine structure of occluded virions, *Virology*, 43, 453, 1971.

123. **Hatta, T. and Francki, R.I.B.,** Similarity in the structure of cytoplasmic polyhedrosis virus, leafhopper A virus and Fiji disease virus particles, *Intervirology*, 18, 203, 1982.

124. **Lewandowski, L.J. and Traynor, B.L.,** Comparison of the structure and polypeptide composition of three double-stranded ribonucleic acid-containing viruses (diplornaviruses): cytoplasmic polyhedrosis virus, wound tumor virus, and reovirus, *J. Virol.*, 10, 1053, 1972.

125. **Martin, M.L., Palmer, E.L., and Middleton, P.J.,** Ultrastructure of infantile gastroenteritis virus, *Virology*, 68, 146, 1975.

126. **Stannard, L.M. and Schoub, B.D.,** Observations on the morphology of two rotaviruses, *J. Gen. Virol.*, 37, 435, 1977.

127. **Bartlett, A.V., Bednarz-Prashad, A.J., DuPont, H.L., and Pickering, L.K.,** Rotavirus gastroenteritis, *Annu. Rev. Med.*, 38, 399, 1987.

128. **Roseto, A., Escaig, J., Delain, E., Cohen, J., and Scherrer, R.,** Structure of rotaviruses as studied by the freeze-drying technique, *Virology*, 98, 471, 1979.

129. **Champsaur, H.,** Les réoviridés: réovirus et rotavirus, in *Virologie médicale*, Maurin, J., Ed., Flammarion Médecine-Sciences, Paris, 1985, 716.

130. **Esparza, J., Gorziglia, M., Gil, F., and Römer, J.,** Multiplication of human rotavirus in cultured cells: an electron microscopic study, *J. Gen. Virol.*, 47, 461, 1980.

131. **Metcalf, P.,** Reoviridae, in *Animal Virus Structure*, Nermut, M.V. and Steven, A.C., Eds., Elsevier, Amsterdam, 1987, 135.

132. **Schiff, L.A. and Fields, B.N.,** Reoviruses and their replication, in *Virology*, 2nd ed., Vol. 2, Fields, B.N. and Knipe, D.M., Eds., Raven Press, New York, 1990, 1275.

133. **Fenner, F., Bachmann, P.A., Gibbs, E.P.J., Murphy, F.A., Studdert, M.J., and White, D.O.,** *Veterinary Virology*, 1st ed., Academic Press, Orlando, FL, 1967, 108.

134. **Nienhaus, T.,** *Viren, Mykoplasmen und Rickettsien*, Eugen Ulmer, Stuttgart, 1985, 48.

135. **Hosaka, Y. and Aizawa, K.,** The fine strcuture of the cytoplasmic-polyhedrosis virus of the silkworm, *Bombyx mori* (Linnaeus), *J. Insect Pathol.*, 6, 53, 1964.

136. **Reddy, D.V.R. and MacLeod, R.,** Polypeptide components of wound tumor virus, *Virology*, 70, 274, 1976.

137. **Hatta, T. and Francki, R.I.B.,** Morphology of Fiji disease virus, *Virology*, 76, 797, 1977.

138. **Frank, H.,** Retroviridae, in *Animal Virus Structure*, Nermut, M.V. and Steven, A.C., Eds., Elsevier, Amsterdam, 1987, 253.

139. **De-Thé, G. and O'Connor, T.E.,** Structure of a murine leukemia virus after disruption with tween-ether and comparison with two myxoviruses, *Virology*, 28, 713, 1966.

140. **Frank, H., Schwarz, H., Graf, T., and Schäfer, W.,** Properties of mouse leukemia viruses. XV. Electron microscopic studies on the organization of Friend leukemia viruses and other mammalian C-type viruses, *Z. Naturforsch.*, 33c, 124, 1978.

141. **Frank, H.,** Lentivirinae, in *Animal Virus Structure*, Nermut, M.V. and Steven, A.C., Eds., Elsevier, Amsterdam, 1987, 295.

142. **Watson, J.D.,** *Molecular Biology of the Gene*, 3rd ed., W.A. Benjamin, Menlo Park, CA, 1976, 673.

143. **Sarkar, N.H.,** Oncovirinae: type B oncovirus, in *Animal Virus Structure*, Nermut, M.V. and Steven, A.C., Eds., Elsevier, Amsterdam, 1987, 257.

144. **Sarkar, N.H. and Moore, D.H.,** Surface structure of mouse mammary tumor virus, *Virology*, 61, 38, 1974.

145. **Bentvelzen, P. and Hilgers, J.,** Murine mammary tumor virus, in *Viral Oncology*, Klein, G., Ed., Raven Press, New York, 1980, 311.

146. **Padgett, F. and Levine, A.S.,** Fine structure of the Rauscher leukemia virus as revealed by incubation in snake venom, *Virology*, 30, 623, 1966.

147. **Nermut, M.V., Frank, H., and Schäfer, W.,** Properties of mouse leukemia viruses. III. Electron microscopic appearance as revealed after conventional preparation techniques as well as freeze-drying and freeze-etching, *Virology*, 49, 345, 1972.

148. **Girard, M. and Hirth, L.,** *Virologie générale et moléculaire*, Doin, Paris, 1980, 352.

149. **Bolognesi, D.P.,** Structural components of RNA tumor viruses, *Adv. Virus Res.*, 19, 315, 1974.

150. **Gallo, R.C. and Montagnier, L.,** AIDS in 1988, *Sci. Am.*, 259(4), 41, 1988.

151. **Montagnier, L.,** Ed., *SIDA — Les Faits, l'Espoir*, Med-Edition, Paris, and Ministère de la Santé Publique et de la Population, Port-au-Prince, Haiti, 1987, 6.

152. **Boyd, J.E. and James, K.,** Human immunodeficiency virus: strategies for protection and therapy, *Microbiol. Sci.*, 5, 300, 1988.

153. **Bolognesi, D.P.,** Human immunodeficiency virus vaccines, *Adv. Virus Res.*, 42, 103, 1993.

154. **Arnold, E. and Arnold, G.F.,** Human immunodeficiency virus structure: implications for antiviral design, *Adv. Virus Res.*, 39, 1, 1991.

155. **Gelderblom, H.R., Gentile, M., Scheidler, A., Özel, M., and Pauli, G.,** Zur Struktur und Funktion bei HIV: Gesichertes, neue Felder und offene Fragen, *AIDS Forsch.*, 8, 231, 1993.

156. **Nermut, M.V., Hockley, D.J., Jowett, J.B.M., Jones, I.M., Garreau, M., and Thomas, D.,** Fullerene-like organisation of HIV gag-protein shell in virus-like particles produced by recombinant baculovirus, *Virology*, 198, 288, 1994.

157. **Ackermann, H.-W. and DuBow, M.S.,** *Viruses of Prokaryotes*, Vol. 2, 1987, 2, 78, 96, and 174.

158. **Luo, C.,** Biochemical Studies on the Structure of Bacteriophage PRD1, an *Escherichia coli* Virus with an Internal Membrane. Ph.D. thesis, University of Helsinki, Finland, 1993, 21.

159. **Finch, J.T., Crowther, R.A., Hendry, D.A., and Struthers, J.K.,** The structure of *Nudaurelia capensis* ß virus: the first example of a capsid with icosahedral surface symmetry T = 4, *J. Gen. Virol.*, 24, 191, 1974.

160. **Stanley, W.M.,** Virus composition and structure — 25 years ago and now, *Fed. Proc.*, 15, 812, 1956.

161. **Franklin, R.E., Klug, A., and Holmes, K.C.,** X-ray diffraction of the structure of tobacco mosaic virus, in *The Nature of Viruses*, Wolstenholme, G.E.W. and Millar, E.C.P., Eds., J. & A. Churchill, London, 1957, 39.

162. **Pollard, E.C.,** The combined effect of thermal and ionizing radiation on viruses, in *Hepatitis Frontiers*, Hartman, F.W., LoGrippo, G.A., Mateer, J.G., and Barron, J., Eds., Little, Brown, Boston, 1957, 355.

163. **Klug, A. and Caspar, D.L.D.,** The structure of small viruses, *Adv. Virus Res.*, 7, 225, 1960.

164. **Caspar, D.L.D.,** Structure and function of regular virus particles, in *Plant Virology*, Corbett, M.K. and Sisler, H.D., Eds., University of Florida Press, Gainesville, FL, 1964, 267.

165. **Offord, R.E.,** Electron microscopic observations on the structure of tobacco rattle virus, *J. Mol. Biol.*, 17, 370, 1966.

166. **Simons, K., Garoff, H., and Helenius, A.,** How an animal virus gets into and out of its host cell, *Sci. Am.*, 246(2), 58, 1982.

167. **Harrison, S.C.,** Virus structure: high-resolution perspectives, *Adv. Virus Res.*, 28, 175, 1983.

168. **Simons, K., Garoff, H., and Helenius, A.,** The glycoproteins of the Semliki Forest virus membrane, in *Membrane Proteins and their Interactions with Lipids*, Vol. 1, Capaldi, R., Ed., Marcel Dekker, New York, 1977, 207.

169. **Coombs, K. and Brown, D.T.,** Topological organization of Sindbis virus capsid protein in isolated nucleo-capsids, *Virus Res.*, 7, 131, 1987.

170. **Horzinek, M. and Mussgay, M.,** Studies on the nucleocapsid of a group A arbovirus, *J. Virol.*, 4, 514, 1969.

171. **Payment, P., Ajdukovic, D., and Pavilanis, V.,** Le virus de la rubéole. I. Morphologie et protéines structurales, *Can. J. Microbiol.*, 21, 703, 1975.

172. **Markham, R.,** Virus nucleic acids, *Adv. Virus Res.*, 1, 315, 1953.

173. **Horne, R.W., Hobart, J.M., and Pasquali-Ronchetti, I.,** A negative staining-carbon film technique for studying viruses in the electron microscope. III. The formation of two-dimensional and three-dimensional crystalline arrays of cowpea chlorotic mottle virus, *J. Ultrastruct. Res.*, 53, 319, 1975.

174. **Rossmann, M.G. and Rueckert, R.R.,** Simple spherical RNA viruses, *Microbiol. Sci.*, 4, 208, 1987.

175. **Harrison, S.C., Olson, A.J., Schutt, C.E., Winkler, F.K., and Bricogne, G.,** Tomato bushy stunt virus at 2.9 Å resolution, *Nature*, 276, 368, 1978.

176. **Liljas, L. and Strandbert, B.,** The structure of satellite tobacco necrosis virus, in *Biological Macromolecules and Assemblies*, Vol. 1, Jurnak, F.A. and McPherson, A., Eds., John Wiley & Sons, New York, 1984, 97.

177. **Cornuet, P.,** *Éléments de virologie végétale*, INRA Editions, Paris, 1987, 37.

178. **Finch, J.T. and Klug, A.,** Arrangement of protein subunits and the distribution of nucleic acids in turnip yellow mosaic virus. II. Electron microscopic studies, *J. Mol. Biol.*, 15, 344, 1966.

179. **Harrison, S.C.,** Virus structure: high-resolution perspectives, *Adv. Virus Res.*, 28, 175, 1983.

180. **Matthews, R.E.F.,** *Plant Virology*, 3rd ed., Academic Press, San Diego, 1991, 127 and 129.

181. **Driedonks, R.A., Krijgsman, P.C.J., and Mellema, J.E.,** Alfalfa mosaic virus protein polymerization, *J. Mol. Biol.*, 113, 123, 1977.

182. **Matthews, R.E.F.,** Portraits of viruses: turnip yellow mosaic virus, *Intervirology*, 15, 121, 1981.

183. **Narang, H.K.,** Scrapie-associated tubulofilamentous particles in scrapie hamsters, *Intervirology*, 34, 105, 1992.

184. **Matthews, R.E.F.,** *Fundamentals of Plant Virology*, Academic Press, San Diego, 1992, 185.

185. **Zyzik, E., Ponsetto, A., Forzani, B., Hele, C., Heermann, K.-H., and Gerlich, W.H.,** Proteins of hepatitis delta virus in serum and liver, in *Hepadna Viruses*, Robinson, W., Koike, K., and Will, H., Eds., Alan R. Liss, New York, 1987, 565.

186. **Ruska, H.,** Ergebnisse der Bakteriophagenforschung und ihre Deutung nach morphologischen Befunden, *Ergeb. Hyg. Bakteriol. Immunforsch. Exp. Ther.*, 25, 437, 1943.

187. **Anderson, T.F., Rappaport, C., and Muscatine, N.A.,** On the structure and osmotic properties of phage particles, in *Le Bactériophage. Premier Colloque International, Royaumont 1952, Ann. Inst. Pasteur*, 84, 5, 1953.

188. **Anderson, T.F.,** The morphology and osmotic properties of bacteriophage systems, *Cold Spring Harbor Symp. Quant. Biol.*, 18, 197, 1953.

189. **Rees, P.J. and Fry, B.A.,** The morphology of staphylococcal bacteriophage K and DNA metabolism in infected *Staphylococcus aureus*, *J. Gen. Virol.*, 53, 293, 1981.

190. **Hotchin, J.E.,** The purification and electron microscopical examination of the structure of staphylococcal bacteriophage K, *J. Gen. Microbiol.*, 10, 250, 1954.

191. **Evans, E.A.,** Bacteriophage as nucleoprotein, *Fed. Proc.*, 15, 827, 1956.

192. **Evans, E.A.,** The comparative chemistry of infective virus particles and their functional activity: T2 and other bacterial viruses, in *The Viruses*, Vol. 1, Burnet, F.M. and Stanley, W.M., Eds., Academic Press, New York, 459, 1959.

193. **Horne, R.W.,** Recent advances in the application of negative staining to the study of virus particles examined in the electron microscope, *Adv. Optical Electron Microsc.*, 6, 227, 1975.

194. **Moustardier, G.,** *Virologie médicale*, 4th ed., Maloine, Paris, 1973, 138.

195. **Eiserling F.A.,** Structure of the T4 virion, in *Bacteriophage T4*, Mathews, C.K., Kutter, E.M., Mosig, G., and Berget, P.B., Eds., American Society for Microbiology, Washington, DC, 1983, 11.

196. **Poglazov, B.F.,** *Morphogenesis of T-Even Bacteriophages*, Monographs in Developmental Biology, Vol. 7, S. Karger, Basel, 1973, 7.

197. **Baschong, W., Aebi, U., Baschong-Prescianotto, C., Dubochet, J., Landmann, L., Kellenberger, E., and Wurtz, M.,** Head structure of bacteriophages T2 and T4, *J. Ultrastruct. Res.*, 99, 189, 1988.

198. **Mosig, G. and Eiserling, F.,** Phage T4 structure and metabolism, in *The Bacteriophages*, Vol. 2, Calendar, R., Ed., Plenum Press, New York, 1988, 521.

199. **Earnshaw, W.C., King, J., Harrison, J., and Eiserling, F.A.,** The structural organization of DNA packaged within the heads of T4 wild-type, isometric, and giant bacteriophages, *Cell*, 14, 559, 1978.

200. **Black, L.W., Newcomb, W.W., Boring, J.W., and Brown, J.C.,** Ion etching of bacteriophage T4 DNA: support for a spiral-folded model of packaged DNA, *Proc. Natl. Acad. Sci. U.S.A.*, 82, 7960, 1985.

201. **Moody, M.F.,** Structure of the sheath of bacteriophage T4. I. Structure of the contracted sheath and polysheath, *J. Mol. Biol.*, 25, 167, 1967.

202. **Simon, L.D. and Anderson, T.F.,** The infection of *Escherichia coli* by T2 and T4 bacteriophages as seen in the electron microscope. I. Attachment and penetration, *Virology*, 32, 279, 1967.

203. **Crowther, R.A., Lenk, E.V., Kikuchi, Y., and King, J.,** Molecular reorganization in the hexagon to star transition of the baseplate of bacteriophage T4, *J. Mol. Biol.*, 116, 489, 1977.

204. **Cummings, D.J., Chapman, V.A., DeLong, S.S., Kusy, A.R., and Stone, K.R.,** Characterization of T-even bacteriophage substructures. I. Tail plates, *J. Virol.*, 6, 545, 1970.

205. **Tikhonenko, A.S.,** *Ultrastructure of Bacterial Viruses*, Plenum Press, New York, 1970, 130.

206. **Eiserling, F.A.,** Bacteriophage structure, in *Comprehensive Virology*, Vol. 13, Fraenkel-Conrat, H. and Wagner, R.R., Eds., Plenum Press, New York, 1979, 543.

207. **Ackermann, H.-W. and Brochu, G.,** Phage-like bacteriocins, in *CRC Handbook of Microbiology*, 2nd ed., Vol. 2, Laskin, A.I. and Lechevalier, H.A., Eds., CRC Press, Boca Raton, FL, 1978, 691.

208. **Ageno, M., Donelli, G., and Guglielmi, F.,** Structure and physico-chemical properties of bacteriophage G. II. The shape and symmetry of the capsid, *Micron*, 4, 376, 1973.

209. **Liljemark, W.F. and Anderson, D.L.,** Structure of *Bacillus* bacteriophage ø25 and ø25 deoxyribonucleic acid, *J. Virol.*, 6, 107, 1970.

210. **Schocher, A.J., Kuhn, H., Schindler, B, Palleroni, N.J., Despreaux, C.W., Boublik, M., and Miller, P.A.,** *Acetobacter* bacteriophage A-1, *Arch. Microbiol.*, 121, 193, 1979.

211. **Walker, D.H. and Anderson, T.F.,** Morphological variants of coliphage P1, *J. Virol.*, 5, 765, 1970.

212. **Goldstein, R., Lengyel, J., Pruss, G., Barrett, K., Calendar, R., and Six, E.,** Head size determination and the morphogenesis of satellite phage P4, *Curr. Topics Microbiol. Immunol.*, 59, 68, 1974.

213. **To, C.M., Eisenstark, A., and Töreci, H.,** Structure of mutator phage Mul of *Escherichia coli*, *J. Ultrastruct. Res.*, 14, 441, 1966.

214. **Pollard, E.,** The action of ionizing radiation on viruses, *Adv. Virus Res.*, 2, 109, 1954.

215. **Eiserling, F.A.,** Bacteriophage structure, in *Comprehensive Virology*, Vol. 13, Fraenkel-Conrat, H. and Wagner, R.R., Eds., Plenum Press, New York, 1979, 543.

216. **Katsura, I.,** Tail assembly and injection, in *Lambda II*, Hendrix, R.W., Roberts, J.W., Stahl, F.W., and Weisberg, R.A., Eds., Cold Spring Harbor Laboratory, Cold Spring Harbor, NY, 1983, 331.

217. **Casjens, S., Hatfull, G., and Hendrix, R.,** Evolution of dsDNA tailed-bacteriophage genomes, *Semin. Virol.*, 3, 383, 1992.

218. **Birdsell, D.C., Hathaway, G.M., and Rutberg, L.,** Characterization of temperate *Bacillus* bacteriophage f105, *J. Virol.*, 4, 264, 1969.

219. **Chiozzotto, A., Coppo, A., Donini, P., and Graziosi, F.,** Icosahedral shape of a temperate phage of *Bacillus megatherium*, *Sci. Rep. Ist. Super. Sanità*, 1, 112, 1961.

220. **Riva, S., Polsinelli, M., and Falaschi, A.,** A new phage for *Bacillus subtilis* with infectious DNA having separable strands, *J. Mol. Biol.*, 35, 347, 1968.

221. **Ackermann, H.-W., Brochu, G., and Emadi Konjin, H.P.,** Classification of *Acinetobacter* phages, *Arch. Virol.*, 135, 345, 1994.

222. **Knopf, U.C.,** Studies on the bacteriophage PS-8 of *Agrobacterium tumefaciens* (Smith and Towsend) Conn: purification and properties, *Arch. Ges. Virusforsch.*, 46, 205, 1974.

223. **Kishko, Ya.G., Ruban, V.I., Tovkach, F.I., Muraschchyk, I.G., and Danileychenko, V.V.,** Structure of *Erwinia carotovora* temperate bacteriophage 59 and its DNA, *J. Virol.*, 46, 1018, 1983.

224. **Takeya, K. and Amako, K.,** The structure of mycobacteriophages, *Virology*, 24, 461, 1964.

225. **Alatossava, T.,** Molecular Biology of *Lactobacillus lactis* bacteriophage LL-H, Ph.D. thesis, University of Oulu, Finland, *Acta Univ. Oulu.*, Ser. A, No. 191, 34, 1987.

226. **Luftig, R. and Haselkorn, R.,** Studies on the structure of blue-green algal virus LPP-1, *Virology*, 34, 664, 1968.

227. **Matsuo-Kato, H., Fujisawa, H., and Minagawa, T.,** Structure and assembly of bacteriophage T3 tails, *Virology*, 109, 157, 1981.

228. **Hendrix, R.W.,** Shape determination in virus assembly: the bacteriophage example, in *Virus Structure and Assembly*, Casjens, S., Ed., Jones and Bartlett, Boston, 1985, 169.

229. **Ackermann, H.-W., Azizbekyan, R.R., Emadi Konjin, H.P., Lecadet, M.-M., Seldin, L., and Yu, M.X.,** New *Bacillus* bacteriophage species, *Arch. Virol.*, 135, 333, 1994.

230. **Kellenberger, E.,** Vegetative bacteriophage and the maturation of the virus particles, *Adv. Virus Res.*, 8, 1, 1961.

231. **Casjens, S. and Hendrix, J.,** Control mechanisms in dsDNA bacteriophage assembly, in *The Bacteriophage*, Vol. 1, Calendar, R., Ed., Plenum Press, New York, 1987, 15.

232. **Ackermann, H.-W., Berthiaume, L., and Kasatiya, S.S.,** Ultrastructure of Vi phages I to VII of *Salmonella typhi*, *Can. J. Microbiol.*, 16, 411, 1970.

233. **Ackermann, H.-W., Berthiaume, L., and Kasatiya, S.S.,** Morphologie des phages de lysotypie de *Salmonella paratyphi* B (schéma de Félix et Callow), *Can. J. Microbiol.*, 18, 77, 1972.

234. **Nutter, R.L., Bullas, L.R., and Schultz, R.L.,** Some properties of five new *Salmonella* bacteriophages, *J. Virol.*, 5, 754, 1970.

235. **Lotz, W. and Mayer, F.,** Isolation and characterization of a bacteriophage tail-like bacteriocin from a strain of *Rhizobium*, *J. Virol.*, 9, 160, 1972.

236. **Ackermann, H.-W. and Gauvreau, L.,** Phages défectifs chez *Chromobacterium*, *Zentralbl. Bakteriol. Parasitenk. Infektionskr. Hyg. Abt. I Orig. A*, 221, 196, 1972.

237. **Leonard, K.R., Kleinschmidt, A.K., and Lake, J.A.,** *Caulobacter crescentus* bacteriophage øCbK: structure and *in vitro* self-assembly of the tail, *J. Mol. Biol.*, 81, 249, 1973.

238. **Keller, B., Dubochet, J., Adrian, M., Maeder, M., Wurtz, M., and Kellenberger, E.,** Length and shape variants of the bacteriophage T4 head: mutations in the scaffolding core genes 68 and 22, *J. Virol.*, 62, 2960, 1988.

239. **Kellenberger, E.,** T4 head assembly: precursors and genetically determined polymorphism, in *Generation of Subcellular Structure*, Proc. First John Innes Symp., Norwich 1973, Markham, R., Ed., Elsevier, Amsterdam, 1973, 62.

INDEX

II. INDIVIDUAL VIRUSES